THE INFINITE

AND

THE FINITE.

By THEOPHILUS PARSONS.

BOSTON:
ROBERTS BROTHERS.
1872.

CAMBRIDGE:
PRESS OF JOHN WILSON AND SON.

CONTENTS.

THE INFINITE AND THE FINITE.

INTRODUCTORY.

WHATEVER is was created, excepting the Creator. If we go up with the ascending series of cause and effect, when we come to the summit we find that which was not caused; for if it were caused, we must go a step farther to find *its* cause. And that which is at the summit, and is itself not caused, has the whole series below it, and must be the cause of all causation.

The totality of being consists of Creator and created. But the Creator, whom we will call God, did not create out of nothing. If God willed that something should be which was not before, His will, His thought, His power, were there, and clothed themselves with the effect and were in the thing He caused to be. He did not create *out of* nothing; for if what He created consisted of nothing, it would be nothing. He created from Himself, and is Himself the substance of all that is; for He created all things from His own substance, and is in all things that are. In the phrase which the Apostles used so often, He is "All in All." Is this Pantheism? It would be so if we held that the totality of created things constituted God, and all the God there is. But then there would be no Creator. The doctrine above expressed is discriminated from Pantheism, by two truths which find no place in that dreary theory. One is, that God, the cause, exists prior to the created universe, or to the effect, and remains as distinct from it after creation as before.

Prior in degree, in nature: I do not say prior in time; for when we attempt a consideration of the beginning *in time* of the putting forth of Divine power in the work of creation, we go beyond the limitations of finite thought. But God is prior to the universe in degree, and perfectly distinct from it while He forms and fills it.

The other truth — which is indeed but another form of the first, or a consequence of it — is that the created universe is as distinct from God as He is distinct from His creation; for God, the Creator, *gives* to every thing He creates to be itself, to be other than Him, and distinct from Him; or, in other words, He so creates every thing that it may possess its own identity or selfhood, and thus be itself and not Him. In the lowest, or mineral kingdom, by force of this selfhood the solid earth with all it contains, without knowledge and without will, performs its functions as the home, the feeder, and the instrument of all things that live. The vegetable kingdom performs its functions, and especially that of drawing from the lower kingdom nutriment, and preparing therefrom nutriment for the kingdom above it. It does this without knowledge and without will, but with the semblance of both in many of what seem its efforts to do its work.

Animals resting on the mineral kingdom and fed by the vegetable kingdom, by force of this selfhood, not only do their appointed work, but do this with a consciousness of every effort they make, and with every effort springing from a will, but with no knowledge whatever of the ultimate purpose of what they do and what they are. And then we rise to man. He is above animals in all respects, and has what they have, and more; and all with a boundless capacity of indefinite development. In nothing is this superiority greater than in the nature and effect of that selfhood or identity which makes him to be himself, and perfectly distinct from his Creator. And the higher character and greater completeness of this selfhood in man is the cause from which the whole of his superiority, including his immortality, proceeds, and the ground on which it rests.

For this universal selfhood of creation is more

than selfhood with man. It is in him ownhood, an ownhood of his life and of himself. For God gives to him to be most truly and most perfectly himself, by giving to him this perfect ownhood of himself and of his life.

The created universe and all things in it are perfectly distinct from God. And man, especially, is himself, and not a part of God, and is most perfectly distinct from God. And now we must remember that distinctness is not independence. For if the universe and man are entirely distinct from God, they are most perfectly, constantly, and absolutely dependent upon Him; just as dependent upon Him always for their continued existence as they were for their creation at the beginning of their being. There can be no time to the eternal, and no space to the infinite; for only to us, the finite and created, do time and space belong. This must be so, however difficult it may be to apprehend and apply this truth. But in the degree in which we are able to comprehend it, and make use of it in our thoughts concerning God, we shall be able to see that creation is not a completed work, but the continual, unceasing, constant, ever-present work of an ever-creating, ever-immanent God.

Then we shall see that not only our whole being, but every part and element of our being, is constantly and unceasingly given to us. As we do not see to-day by the light which came to us yesterday, so we do not live to-day by the life which came to us yesterday or at our birth. We see always by the light which comes now, and we live always by the life which comes now. Every affection, every motive, every feeling, every thought or other intellectual state or act, comes to us from the Source of life, then when we possess it.

But because man has life, and this ownhood of his life, he has freedom. I shall have occasion to say more about freedom hereafter. There are those who deny it altogether. The argument for its impossibility, which to some minds seems decisive and unanswerable, is this: Whatever is must be the effect of some cause. That cause could produce only that effect, and must produce that effect. But that cause is itself the effect of a prior cause. Carry the series as far backward or forward as we will, the law remains the same. Always whatever is must be the effect of a cause which could not but produce that effect, and could not produce any other effect. This argument, which is not wholly unsound as to things without life, is fatally erroneous as to things having life, because it takes no notice of a higher law than that on which it rests; and this is that freedom is an element of life and belongs necessarily to life. An organism which is wholly without freedom has no life; it is a machine; it is dead. The possession of freedom of some kind is that which differentiates living organisms from machines. Indeed a living organism when it loses all its freedom loses all its life, and becomes less than a machine, because it cannot be moved by a force from without. Freedom may be variously modified in things having life, but cannot be wholly suppressed and leave any life. And this necessary element of life, this freedom which belongs to life, enables things having life to act in some way and measure on things not having life, and thus affects the series of cause and effect which belongs to them.

Man's ownhood of his life, and his freedom, are as perfect as if they were given him at the moment of his birth, to last while he lives and be always perfect. This freedom, as to the spiritual part of his nature and his spiritual destiny, is always perfect. As to his external life and nature, it is constantly modified and restrained just so far as his spiritual interests require. But his spiritual freedom is constant. Into this ownhood, this freedom, all that his Creator gives him is constantly given. Life, and all the moral and spiritual elements of his life, are subjected to his own will; and he may do with them what he wills. He may use them as they were given to be used, or he may pervert and misuse them.

Hence man, and man alone, has duty. Hence he determines his own destiny, and he does this by determining his own character. Hence are for him the hope and the possibility of eternal happiness. And hence this is only hope and possibility, and not certainty. In later pages I may endeavor to show that the highest conceivable happiness of a created being is to receive into his own will and by his own choice, and in his own freedom appropriate to himself, divine life and the happiness which belongs to it; and because this is the highest happiness, the divine goodness, because perfect, must so create and govern him as that he may be able to do this and thus enjoy this happiness; and to this end must give him ownhood of his life, and freedom; and freedom, if real and actual, must be capable of abuse, and equally capable of use and of abuse.

If we return to the material universe, there too we shall find that creation is a constant work; and that the support, the maintenance in being of what has been created, is a continual creation of it. All the forces and energies of Nature are but forms and activities of the Divine force which forms and fills them, and which is determined in its action and manifestation and effect by the inmost form or nature of the created thing in which and through which it acts; and the methods of its activity are what we call the laws of Nature.

WHAT IS MATTER?

We say, and we say truly, we live in a material body, and in a material world. But what is matter? It has been already said that the Creator creates the world from Himself, from His own substance, by causing that to flow forth somewhat as heat and light flow from the sun and sustain the earths around it. This effluence, when it reaches the farthest limit to which it goes or needs to go, forms and constitutes the substance of the material body with its organs and sensories, and of the material world. The infinite wisdom from which it flows forms the material world, and forms all the things of the material world in such perfect adaptation to our organs and sensories, and it forms our organs and sensories in such perfect adaptation to our minds, that through these organs and sensories the outer world affects the mind in such a way as to produce sensations and what we call perceptions. And all this adaptation of the world without to the sensories and the senses, and to the mind which is within the senses and receives their report, is so perfect, so wonderful, as to offer an irresistible proof that intelligent design has presided over creation.

But what do we know of matter? It is certain,

and to whoever thinks it is obvious, that we *know* directly only our own sensations, and our thoughts about them. All the rest is inference.

You are looking at a tree. What is it that you, the perceiving being, see? Only a minute picture painted on the back of the eye. It is painted there because the eye is an optical instrument, which, in accordance with the laws of optics, brings all the rays of light from that tree to a focus on the retina, just as the glasses of a photographic camera bring all the rays of light from an object to a focus on the plate, and paint its picture there. This is all the mind can see, if indeed it can be said to see this.

Why, then, do I think I am looking at a tree of that size and in that place? Certainly not from sight alone. To the babe the moon is as near as the lamp on the table. Presently he begins to touch and handle things. He moves, and reaches some things sooner than others. The sense or thought of time, of space, of place, shape and dimension, of effort and resistance, dawns upon him. Day by day his sensations grow more numerous and diversified. Unconsciously he compares them and draws his inferences; and all this before he is old enough to know his own thoughts distinctly, or remember them afterwards. So he grows up with a world around him, of which he has no more doubt than of his own existence. Nor need he or should he doubt. It exists as much as he does. He thinks he knows that it is there where he sees and feels it, and that it is what he sees and feels. But all he does know, or can know, is that through his sensations, and his thoughts about his sensations, *something* produces the impressions which he has.

Idealist philosophers in all ages have followed out this train of thought, until it led some of them still farther, — even to the conclusion that there was no real world outside of sensation and of thought. This was a fatal error.

Their argument, that sensation can tell us only of itself, and that thought about sensation can tell us only that these sensations have a cause, is sound. But their conclusion, that nothing without us causes our sensations, so offends our inevitable convictions, that these philosophers never believed it themselves, and none ever believed it. I cannot be more sure there is a ME, than I am that there is a NOT ME. And it is true; for the totality of being consists of God, who is Being in Himself and the Source of all being, of spiritual substance created by Him, and of material substance created by Him; and this material substance is created for the wants and the use of spiritual being, and is adapted to those wants and that use. But of what this substance is, *in itself*, and independently of the sensations it produces, we know absolutely nothing. The statement that "mind makes matter," if used in one sense, has some truth; and in another has none whatever. It is inaccurate if when we say "make" we mean "create." The mind does not create any thing: there is but one Creator, — God. That which underlies matter and is its substance flows forth from the Divine substance. But mind causes, or rather co-operates to cause, the form, shape, incidents, and appearance, under which we give to this substance the name of Matter. This is not a false appearance; it is not an appearance as of something where there is nothing. It is a true appearance of a most actual reality. But this reality is one of which we can know nothing whatever, but from the action of mind on the impressions made by this substance on the mind through and by means of the senses. Nor can we in this way learn or know any thing of this substance as it is in itself.

The exceeding importance of this truth lies in the rational belief which it permits us, that we may have a body, a home, and a world, when we leave this world. If material substance is but the effluence from and of the Divine substance, caused to be and to affect us in a certain way for our use and benefit while we live in this world, the same effluence from the same Divine substance may provide for us, when we rise from this body, a body and all organs and sensories of spiritual substance, and a home and a world of spiritual substance, as exquisitely adapted to our organs and sensories, and to act on us and on our minds through our organs and sensories, as this world now is.

We have no reason to believe that we, our bodies, or our world, will be just the same then and there that they are now and here. But I would defer to a later part of this work my reasons for believing that they will be not the same, but similar, or rather correspondent. Let me now, however, add a word concerning Space and Time. They are not *beings*, not existing things: they are instruments or forms or methods of thought. And in saying this I only repeat what all philosophy now asserts; and I will not attempt to present the familiar argument which leads to this conclusion.

We look at an insect one-tenth of an inch long. We wish to examine it, and we make use of a lens magnifying tenfold in linear direction, and the insect is now an inch long. We say it *is* one-tenth of an inch long, and *seems* to be an inch long because we see it through that convex glass. But it is the convex lenses of the eye which make it appear one-tenth of an inch long. If they were more or less convex, it would be longer or shorter. Perhaps we see, in a book on Entomology, two drawings of the insect, one a tenth of an inch long, and this is marked "actual size"; and the other an inch long, and this is marked "magnified ten times." And yet one of these is just as much the actual size as the other. Seen through the lenses of the eye, it has the shorter length; if we add another lens, it has the greater length. If the added lens is concave, it looks less than one-tenth of an inch. We cannot see it at all except through the lenses of the eye; and their form and arrangement, and nothing else whatever, determine the length it shall seem to have. This we may diminish or enlarge at pleasure by using additional lenses; and through all these lenses, those in the head and those outside of it, light, acting in accordance with certain known laws, paints the picture we contemplate, and paints it such in shape and dimension as those lenses — whether made for us or made by us — determine. What, then, is an inch, or a tenth of an inch, in length?

Let a reader imagine that, while he sleeps to-night, the whole world, with all things in it, becomes ten or a hundred times larger than before; and that he too, and all that is in him or belongs to him, within or without, and all motion and all action, grow in precisely the same way and proportion. He could know nothing of this change when he wakes. Measured by any standard he could apply, spontaneously or by effort, the outer world would be just what it was before, and would seem so.

Has it, then, any actual size or place? It has *existence*, but it has no size or place excepting what sensation and thought cause it to have.*

---

* One of the latest and most important discoveries of natural science has suggested that all the forces and energies of Nature, electric, magnetic, or actinic, heat and light, and even motion itself, are resolvable into each other, and are but the varied forms and effects of one force. This new truth is sometimes called the Doctrine of the Unity of Force. It is not yet perhaps established, and certainly not well understood. When it is, it will be extended upwards until it embraces mind and thought and affection; and yet farther, until it covers all the life and activities of the spiritual world; and yet farther, until it touches the throne of God. For there is primarily and originally but one Force; and it is the Infinite Love of God which — in itself utterly inaccessible to defining thought, and beyond all adequate conception — comes forth, self-impelled, and creates and fills all forms of being. If, then, we ask why, if all forces are one in essence and in origin, they are so diversified in their appearance and in their action, we find the answer in the principle that this One force or energy is always determined in its manifestation, its character, its aspect and effect, by the inmost nature of the things which it forms and fills; which it forms for that purpose and fills with that effect.

The oldest religious philosophy which ever existed among men, so far as we know, had its home in India. The foundation-myth of the whole was, that, when Brahm proposed to himself to create the universe, he took as his wife Mayá, and all things are their children. Brahm is the name of the absolute and infinite Being, from whom

Then, What and Where is the external world we now live in? It is what the mind makes it and where the mind places it, acting on impressions made upon the mind through the senses by material substance.

And What and Where is the external world of the life after death? It is what the mind makes it and where the mind places it, acting on impressions made upon the mind through the senses by spiritual substance.

And what are material substance and spiritual

---

all being proceeds. Maya means "appearance." This, at least, from all I have been able to learn of it, was its primary meaning. This name is sometimes translated "illusion"; and if this be its true meaning, it would follow that the falsity of a pure idealism existed even then. But if, or while, the word meant Appearance, we may believe that they thought that all existing things received their being from the Infinite, and their form and manner of being from Appearance. Then, they who, after the fashion of that age, expressed their philosophical theory by that myth, held that whatever we see or hear or feel or taste or touch is but the effect, manifestation, and appearance of that of which we can know nothing else from our sensations, or from sensuous thought concerning our sensations, than this Appearance; and the outer world is made to bear this appearance to our senses, and to our minds through our senses, that we may thus have a body, a home, and a world. This we believe; and also that the need of these will go with us when we leave this world, and will be supplied as it is now. And all external things will be seen and known in Space and Time; but Space and Time so modified and changed, and with them all external things, as our new condition will then require.

substance? Two forms or modes of the one substance which flows forth from its Divine origin, and takes one or the other of these forms, to provide for man a body and a world, first in this beginning of life, and then in the life which follows this.

As God is the beginning of all being, so the knowledge of God is the beginning of all knowledge concerning things which belong to man as immortal. Let us now inquire into the origin and nature of man's belief of God, and the manner in which and the purpose for which this belief is given to us.

I. OF THE BELIEF IN GOD.

It has been often said that a belief in God, of some kind or measure, belongs to human nature, and is inevitable and universal. If any tribes are found so utterly savage that they seem to have no religion whatever, it is said that a more thorough knowledge or a more careful inquiry would detect some notion of a governing God, although it might be dim and clouded. And if men of ability and culture deduce their unbelief from what they consider logical reasoning and declare it unreservedly, it is then said that they are mistaken as to their own opinions, and that within their ratiocinative denial lies an unconscious belief, because the heart refuses to listen to argument on this point. It may be doubted whether this be quite so.

There seem to be some savages so wholly brutified that no spark of religion, even in the perverted form of superstition, can be detected by the sharpest scrutiny.

On the other hand, there have been in all ages, and certainly are in this, men distinguished for intellectual power and cultivation, who not only assert, but manifest, in the whole course of their thought and the whole structure of their opinions, absence of all belief in a God. This absence must be admitted as to them, unless we are prepared to say that this belief is so necessary a part of human thought that no evidence can prove its absence. We may at least conclude that if a belief in the existence of a God cannot be wholly extirpated from a human mind, it may be to the last degree dim and feeble in such savages on the one hand, and in such men of high culture on the other.

How is this to be accounted for? Here are two classes of men standing at the opposite extremes of intellectual ability, natural or acquired; and they agree in this most exceptional condition of mind, — a condition so exceptional that many doubt whether it can exist, and would not admit its existence without decisive evidence.

If the class of nearly brutal savages stood alone, there would be no difficulty. It would be said only that such an instance proved that something more than mere human nature was necessary to this belief; or, in other words, that it was possible for beings to exist who were no higher than animals in this respect, but in others lifted so distinctly above animals that the name of man could not be refused to them.

But how is it when we find these savages stand in this respect on a common ground with some who are commonly ranked among the ablest thinkers that have lived? What is the connecting link between them? Or if we go farther, and say that we find such instances not only at these extremes of intellect, but that all the way from the bottom to the summit there may be found some — a very few, but still some — of every grade of power and culture, who stand in this exceptional condition of mind, the question then takes this form: what is the connecting link which gathers into this class of persons who deny or do not know that there is a God, men who in all other respects are so entirely apart, and so far apart?

The answer we would give requires some analysis of the human mind and character.

II. OF THE NATURAL INTELLECTUAL FACULTIES.

We often say that man has a body and a soul, and only these; and whatever belongs to him as a part of his selfhood belongs either to his body or to his soul. If we accept this as an accurate and exhaustive division of the elements of human nature, let us look a little at each of them.

Man's body makes it possible for him to live on earth. Not only so, but his material body is exquisitely adapted to this material earth. The indefinitely various bodies of all animals are adapted to the earth, so far at least that it provides for them a home, where the individual existence may be supported and preserved, the race propagated, and some degree of comfort or pleasure be enjoyed. But animals differ among themselves exceedingly on this point. Some get vastly more out of the earth than others. Some are adapted to the earth far more completely than others, so that they can utilize it to a far greater extent. On this point, as on most others, man far exceeds all other animals. Not, however, because his body is so much better adapted to the earth than theirs, but because he has also mental faculties additional to theirs; or, if the same in character as theirs, of far greater strength, which are exquisitely adapted to bring the general adaptation of the body to the earth into the fullest development, the greatest activity, and the most important results.

Thus man alone can make and use tools or instruments. By these he asserts a superiority over all animals, and a far greater capacity than theirs for making use of the earth. But beyond this, indefinitely beyond this, he has faculties perfectly adapted to observation of the earth, to an investigation into all its forms and forces, its laws and their results and manifestations. Of this nothing is within the reach of animals. But man possesses powers of this kind, of which we may not say there is no limit to them or to their work, but may say that, far as we have advanced in this direction, every step forward has revealed new possibilities, and given new cause for effort and for hope.

It is by the cultivation of these faculties that all natural science has reached its present stage of growth. Among them may be ranked even the mathematical faculty. This, in theory and in some of its working, seems lifted above the earth. It deals with abstract quantities and relations, which exist only for the imagination. But the laws which it investigates are only the inner laws of material things. When the mathematician has pushed his calculus to the farthest result he can now reach, and has learned all it can tell him of the most transcendental curves and motions, the astronomer takes these results and applies them to the sky, sure that the deductions of pure reason are one with the laws of the universe. Why are these truths within these laws as their guides and masters? Only by anticipating what might perhaps be better said at a later period, can we answer this question: they are there because the reason which placed them there and works within them is one with the reason in man which finds them there.

Nor is it astronomy alone which seeks the aid of mathematics. Chemistry, botany, and even geology, are all invoking it. And the science which bears the name of applied mathematics brings down the best work of the pure reason to render service to man in carrying on the business of life and procuring its enjoyments.

How much farther man may go in these directions none may say. It is only certain that we have not reached the end, nor yet come within sight of the end. The impossibilities of a few generations ago are our actualities. Their dreams belong to our waking life; and, in some things now familiar to our use, we have gone beyond even the dreams of former generations. All this excites, and perhaps it justifies, a boundless hope for the future: the hope of a future in which the material universe, and all there is in it of substance, or force, or activity, will in some way minister to the wants and promote the happiness of man.

But of all these faculties, even in their widest possible or conceivable development, one thing remains to be said.

None of them, no one of them nor the whole together, belong to man as an immortal being, or necessarily imply, or lead the thought to, immortality. And if they alone belonged to the human mind, and constituted the whole of its intellectual power or wealth, they could never lead the thoughts towards any life other than life on this earth.

We have indeed at the present time what may be regarded as positive proof, that these faculties, although exerted with the utmost vigor and reaching the highest attainable success, have, of themselves, no power and no tendency to look beyond this world. We have this proof in the fact that so many persons who stand in the very first class of scientists at this day have but little scruple — in some instances, none whatever — in letting it be understood that they consider the things of another world, or the fact of another world, as not within the scope of human knowledge or rational belief.

Some evidence to the same effect may be found in the fact, that other eminent scientists, who declare their belief in God and in religion, separate this belief entirely from scientific inquiry or thought; and some — as Faraday, for example — are of opinion that religious faith is to be protected and preserved only by an entire severance from those other topics or processes of thought and knowledge.

The self-contentment of these human faculties in their own work has led to a conclusion which seems to be now quite widely held, — that the difference between man and the lower animals is a difference of *degree* and not of *kind;* inasmuch as we can see in the animal kingdom the beginnings of all those faculties which grow into greater strength and do far greater work in man. This is true in respect to all the faculties adapted only to utilize the earth for practical or scientific purposes. But there is still a decisive difference between man and animals, even in respect to these faculties. Every animal is born into possession of the faculties his parents had, and can never enlarge them. He is not educated into the use of his faculties; nor does he inherit any advance in the use of them if his parents made any. In fact they made and could make no advance. The instances known of breeds of domestic animals acquiring and transmitting new powers or new methods of using common powers, are, in the first place, too slight to form a noticeable exception to the rule. And in the next place, whatever may be supposed by some theorists, we have no evidence that even those improvements are ever effected except among domestic animals, or are ever the results of mere animal effort unaided by the care and training of men. And even if we admit a gradual enlargement and gain in successive generations of animals, produced by "the struggle for life," or "the survival of the fittest," or other causes, one thing remains true and certain. It is, that animals cannot voluntarily and consciously, and by efforts made for that purpose, transmit all that one generation has learned to the next, and so by accumulation through ages add to the constantly growing mass of knowledge of the external universe and to their ability to make use of this universe. And so we reach the conclusion, that man has all the faculties to utilize the earth which animals have, and yet other faculties which make his power to utilize the earth vastly greater than that which animals possess.

III. OF THE SPIRITUAL INTELLECTUAL FACULTIES.

All the faculties we have spoken of above are natural to man. We use this phrase because by Nature

— from *nascor*, to be born — we mean the world into which man is born and all that belongs to it, and all those faculties of his mind which enable him to study and comprehend the external universe, and enlarge indefinitely his scientific knowledge thereof, and make use of its substances, its forces and its laws, for his own benefit. We call these faculties, therefore, his *natural* faculties. It is a convenient phrase; and, if the definition we have given be remembered, we shall not be misunderstood when we make use of it.

The important fact which we would now remember is, that the natural faculties of man do not lead him to any knowledge or belief of God or of another life; and do not, of themselves, cause or suggest any thought of that kind. They are adapted to this earth in the most perfect manner. This is their end and purpose; and to this end and purpose they are entirely adequate. And to this end and purpose they are confined and limited, in their capacity and in their action. Of themselves and in their own activity, they stop short of any thought or any knowledge or belief, other than belong to this world and the things of this life.

And yet it is certain that all or very nearly all men have a belief, or thoughts, referring in some form or measure to another life. How is this to be accounted for? By the fact that man has, in addition to his natural faculties, another class of faculties. These differ so entirely from the natural faculties, that they constitute between man who possesses them, and animals who do not possess them in the slightest measure, a difference which is not a difference of degree, but radically and essentially a difference of kind. This difference constitutes indeed the essential difference between man and the lower animals. Only by considering it, only by learning as well as we may what this difference is, and whence it arises, and what is its effect, can we arrive at just results concerning human nature.

It must be very difficult to many to see that this is so; and to some it must be impossible. The reason is, that it can be seen to be true only by an exercise of these higher faculties, because they only can take cognizance of this matter. And this again is a hard saying. Perhaps it may be made easier by some considerations which may lead us to see why it is that man possesses faculties of these two classes, distinguished from each other in this way.

I must, in the first place, assume that there is a God; and this I do without the slightest attempt, at this time, to prove His existence, and still remembering that the natural faculties cannot recognize this truth.

In the idea of God, and by this I mean in every actual idea of God, must be included the idea that He in some way made the universe and man, and governs both. At all events, I assume this; and admit at once that if this be not admitted I have no basis whatever for the argument I now attempt to present.

It does not necessarily follow from the fact that God made man, that man is immortal. But if it be admitted that God made man and the universe, and placed man here and made man such as he is, there are reasons enough derivable from man's relations to the universe, and from his mental, moral, and physical nature, for believing that man is immortal. But these reasons can have no force or effect unless they are seen and weighed by the spiritual faculties. It is my present purpose to show the reasonableness of believing the existence of these faculties; and I will not attempt to show this by an inference from the work they do, but again assume man's immortality.

I stand then here: I assume that God made man, and made him to be immortal. This means that God made the universe as a home for man, not only in the beginning of his being, but made him such, that, after he had ceased to live in this world or state or manner, he should live in another world or state or manner, and should live there always.

And now my argument is, that, if God made man to live first in this world, he would give him faculties adequate to this world and suited to make the most of it. And if God made man to live another life after this has ceased, he would make this life such, in its relation to the other, that it might be preparatory for the other, or helpful to man considered as one who would live after he ceased to live here; — and would give him faculties by which he would make this use of this life.

We know that man has faculties, many and various faculties, perfectly adapted to utilize this earth in every way that begins and ends upon this earth. So much no one ever doubted. But to him who believes that man only *begins* to live here, and that God places him here so to begin life, and gives him these faculties adapted to this beginning of life, — to him it will seem most reasonable to believe that God made him to begin life here, *with a view* to his living hereafter, and with the purpose that by living here he might live more happily hereafter; and that if God gave man faculties admirably adapted to utilize this life *in itself*, so he would give him faculties by which while living here he could utilize this life *for another life.*

More than this. If life is limited here, and the life hereafter is without limit, it must be infinitely more reasonable to believe that God would give man faculties suited to prepare him by a due exercise thereof for that unending life, than that he would give him faculties suited to live comfortably here.

And because that life cannot be precisely the same with this, but may be similar, and, if this life is intended to be preparatory for that and as an apprenticeship or school for that, must be believed to be in important respects similar; so it is reasonable to believe that the faculties adapted to utilize this life as a preparation for the other life are not the same, but are in important respects similar to the faculties which are adapted to this life in and by itself.

And one point of similarity between these two classes of faculties may reasonably be believed to be this: as the faculties possessed by man, to utilize this world for all its possible advantage to him while living here, do not benefit him excepting in proportion as he makes a suitable use and exercise of them; so the faculties possessed by him that while living here he may prepare himself for greater happiness hereafter, do not benefit him in that respect excepting in proportion as he makes a suitable use and exercise of them.

It is by these higher faculties that man thinks of

another life, and of a God who is the creator, governor, and preserver of all things. It is by these faculties that man thinks of the things of religion; that he thinks spiritual thoughts. And we give to these faculties the name of spiritual faculties.

We have spoken of the natural faculties and spiritual faculties as entirely different and distinct. It might be more satisfactory to some minds, if we said only that they worked in entirely different ways, on entirely different planes, and with entirely different results. Perception, comparison, imagination, reason, analysis and synthesis, are all employed as natural faculties; and they are all employed as spiritual faculties. We know so little yet of the forces of the external universe and the way they run into one another and manifest a unity in the midst of diversity; and so much less of the spiritual faculties and their true relation to the natural faculties, that we are not prepared to see distinctly how far they are separate and distinct, in themselves, or how far they are the same, differing only in the modes of their activity; or differing as in degree or rank, the one set being lower and the other higher, or the one set external and the other internal. Without pausing now on this point, the difference between them is abundantly sufficient to prevent our being misunderstood when we speak of the faculties last described as spiritual faculties.

The radical difference between them is this. By the natural faculties, man may investigate and master the external universe. But they go no further; not even in thought, or desire, or hope. They would be and do just what they are and do; and we can suppose that they might have gained for man all the treasures they have won; and that they might gain for him all that ever shall be won in the same direction if the wildest dreams of scientific enthusiasm become realities, — whether man lives after death, or is extinguished when he dies. This could make no difference to them, none to their work or its results. They would be and do all they now are and do, if this external universe, born of itself, had come into being by its own energies, and was in its complex God and the only God.

The utter unreasonableness, the fearful absurdity of such a supposition can be apparent to, can be even suggested by, only some exercise of the spiritual faculties. For it is to them and to them only that we owe whatever we have of thought or hope or belief of an immortal life, and of a personal Father of us and all things whom we may obey and trust and love. We should not have the possibility of any thoughts like these had we no spiritual faculties. We cannot have the actuality of any such thought or faith, if these faculties are not awakened and exercised.

In another respect these two classes of faculties are perfectly alike. While both are necessary elements of human nature and must belong to man because he is man, they may, both or either, exist only potentially, and not actively. Or they may have any degree of activity; the natural faculties from the least and lowest found in the most savage life, to the greatest height they have yet ascended or ever may ascend, in the most powerful and cultivated intellects; the spiritual faculties from the total or almost total absence of any spiritual thought, through all degrees of dimness or obscurity or clearness and strength of such thought, to the most intense religious conviction. So, too, they may, both or either, be misguided and mistaken. The brutal life of the savage, who makes little or no use of his natural faculties, finds its exact correlative in the ignorance or unbelief of those who make no use of their spiritual faculties, whatever they may do with their natural faculties. The wild and foolish and sometimes mischievous theories and suppositions which have prevailed as to external things, in different places and ages, find their exact correlative in the fantastic and irrational and sometimes most mischievous theories and suppositions which have prevailed as to matters of religion.

But, while these two classes of faculties are alike in these respects, they manifest their distinction from each other by their perfect independence of each other. There is no form and no degree of disregard or even contempt or dislike of the natural faculties and their exercise, which may not coexist with some form of sincere and profound religious faith. Asceticism proved this; and perhaps all tendency to asceticism is a tendency in this direction. On the other hand, no kind and no degree of vigor, activity, and success of the natural faculties implies, or of itself indicates, any action of or any vitality in the spiritual faculties.

This law is at once illustrated by, and gives us an explanation of, the fact to which reference was made at the beginning of this essay; the fact, namely, that to-day utter unbelief of religion is to be found mainly in those two classes of human beings who stand, as to the exercise of the natural faculties, at the opposite poles of human nature; surrounded equally in both cases, as to the spiritual faculties, with arctic barrenness, silence and death. At the one extreme stands the lowest savage; utterly wanting in spiritual thought and belief, because, while his natural faculties are only so far brought into play as to preserve his life and perpetuate his race, his spiritual faculties are, if possible, still less active. At the other extreme stands the eminent scientist, who, as to his spiritual faculties, and all that they would teach, is just where the savage is; because his natural faculties are exercised with an intensity, and the exercise itself and its results are loved with a passionate and exclusive devotion, which leaves nothing of interest, nothing of power, nothing of life, to his spiritual faculties. And he knows nothing and believes nothing of what they would tell him. In one half of his nature, how far beyond the savage! in the other half, how entirely the same!

While writing this, I met with a paper in a recent English periodical, by Professor Huxley. I need not say that this gentleman stands high in the highest class of scientific men. In this paper he is reviewing a book in which the *ends* of creation and the origin of life, and other topics of like kind, are much considered. Mr. Huxley, after noticing the author's views and some of an opposite kind, comes to the conclusion that all such inquiries are vain and fruitless. He ends thus: "Why trouble one's self about matters which are out of reach, when the working of the mechanism itself, which is of infinite practical importance, affords scope for all our energies." In this short sentence four things are worthy of notice. The first is, that he regards all truth of the kind which the author he is reviewing seeks — that is, all spiritual truth — as "out of our reach." The second, that he regards the universe as "a mechanism." The third, that the working of this mechanism considered *by itself* is a topic "of infinite importance." The fourth, that the study of it, its laws, forces, aspects and relations as a mechanism, "offers scope for all our energies."

And, on his own ground, he is perfectly right on all these points. He has used the natural faculties only, — used them with earnestness and eminent success, but used only them. He knows no other, and believes in no other. And it is entirely true that all the topics which he thus contemns are "beyond the reach of those faculties;" that, so far as they can see or teach, the universe is "a mechanism;" that the investigation of it as such is the only work which those faculties can recognize, and to them it is "infinitely important;" and it "affords scope for all their energies," and is their proper object, and suffices and ever will suffice to employ and to exhaust all their strength.

Speak to one who has devoted his life to the employment and invigoration of these lower faculties, while the higher slept in utter inaction and unconsciousness, — speak to him, in any words which man can use, of God, and another life which will not end, and of faculties which are given us that we may learn *truth* concerning these spiritual things, and utilize this truth for advantages which are related to all that the lower faculties can give as eternity to time, and how is it possible that he should understand you any more than if you spoke in a language he did not know?

As we can have no idea whatever of any religious truth except by the exercise of these higher spiritual faculties, so the religious ideas we have will be distinct or dim, strong or weak, constant and habitual, or intermitting and infrequent, precisely as these faculties act distinctly or obscurely, vigorously or feebly, continually or with longer or shorter intervals of slumber. So, too, these ideas will be accurate or mistaken accordingly as these faculties are wisely or unwisely exercised.

These faculties are given us that we may be prepared by a life in this world for life in another world. The lower, natural faculties relate to this world only. It is therefore obvious that an undue love of this world, or worldliness (and so it should be called whatever form it takes), must tend to give undue development and strength to the natural faculties, and in the same degree to suppress the spiritual faculties. And most true it is that there are none in whom this worldliness does not exert some influence adverse to that of the spiritual faculties; and few in whom this adverse influence is not powerful; while in many persons it is so strong and so constant, that if it does not wholly prevent any exercise of the spiritual faculties, — that is, if it does not wholly prevent any thought or belief of God or immortality, — it makes such thoughts poor, feeble, unfrequent, and inoperative.

It may be what would be called the cares of this world, the pursuit of wealth, power, or fame; or it may be only the labor and devotion to daily occupation which seem necessary for comfortable subsistence; or it may be an extreme sensuous enjoyment of what the world offers to the senses; or it may be the engrossing love of external knowledge or science for its own sake or for its rewards; — it may be either or any of these causes, which confine our thoughts and concentrate our interest upon things which belong to this world only. Then, in the same measure, the faculties we possess for the sake of the other life and which refer to that or to a preparation for that, are torpid and powerless. Let me again suppose that we live in this short life that we may prepare in it for another life of indefinite length, and that our life through the unending hereafter will be profoundly affected by our life here and by every part of it; then let us cast the light of these truths upon the course of belief, affection, and conduct which prevail in society, and what must be our conclusion?

IV. OF THE NATURAL AND THE SPIRITUAL AFFECTIONAL FACULTIES.

The same distinction which exists between the natural and the spiritual faculties considered intellectually, exists between them if they are considered affectionally.

The man, as to all within the body, consists of will and understanding. All his powers and all their functions may be referred to what he thinks, believes, or denies, or else to what he feels, loves, or hates. Because we begin our existence in this world, our natural intellectual faculties are adapted to this world. But we have also natural affectional faculties; and these, like the natural intellectual faculties, are adapted to this world, and to all its gifts, its requirements, and its relations. But as we live in this world to prepare for another world, we have not only spiritual intellectual faculties by means of which this preparation may be made, but also spiritual affectional faculties to co-operate with them, or rather to lead, in this work of preparation.

We have seen that the natural intellectual faculties may have any measure of strength or culture, or activity or success, and yet remain purely natural, with no light cast upon them or on their work by the spiritual intellectual faculties. Precisely so the natural affections may take any form or any character, be strong or weak, be good or bad, and either good or bad in any degree, and yet remain only natural affections, with no life in them from spiritual affections.

There can be no enjoyment of life in this world, without mutual kindness. This emotion all good men feel, and all wise men cultivate. So, too, the race could not be perpetuated if there were no parental affection. The utter feebleness of the infant and his perfect dependence upon his parents would make the continuance of his life impossible without their care; and this care would be impossible without parental affection. This love is therefore universal. If it ceased to exist and to be powerful in any race, that race must die out.

We may go to a yet higher love, — that which unites husband and wife. The sexes must come together, or the race will not be perpetuated. This end might be answered without choice, constancy, or permanence in this relation. But these add infinitely to the comfort and enjoyment of life. The great majority of men acknowledge this in theory; and great numbers live in the practice of it.

Is not all this good? Certainly. But that is not now the question. It is, Are these affections natural or spiritual? As there are natural intellectual faculties and spiritual intellectual faculties, and natural affections and spiritual affections, so there is natural goodness and spiritual goodness. And to the question, Is not all this good? while we answer, Certainly it is, we may then ask the farther question, Is it natural goodness or spiritual goodness? And to this question we must apply the same test as before. The natural affections look to this life and to this alone. The spiritual affections look to God and to another life. The natural affections, born in this world, live here and die here. Spiritual affections, born in this world, rise above it, and never die. The natural affections belong to man as he is one among the animals; and the beginnings or rudiments of them, as of the natural intellectual qualities, may be found in his brother animals. But to find the first trace or semblance of spiritual affections we must go above animals: we must go to man; we must go to that in man which makes him other than animals. We must go to man as an immortal being, and to that part of his nature which is given him to the end that, while he is living the temporary life he shares with animals, he may be preparing for immortal life. And now we have the means of answering the question, What is this goodness?

If it be only as above described, it is only natural goodness. Later in this work it will be my endeavor to show that these two classes of faculties — natural and spiritual — stand in such relation to each other, that the highest possible culture of the natural intellectual faculties prepares them to be the most effectual instruments of the spiritual intellectual faculties, when these take their proper place and do their proper work; for then they will not renounce what the natural intellectual faculties have gained, but will find in all of it inexhaustible illustration and indispensable assistance. Just so, when the spiritual affectional faculties command the character, they will not renounce natural goodness, but will rejoicingly adopt all of it, and make it their own form and expression, and fill it with new and a far more vigorous life.

The distinction between the natural and the spiritual intellectual faculties is much more clearly defined and more easily seen than that between natural and spiritual affections. But this too is visible, or discoverable; for we have the unfailing test of a reference to God and immortality in the one class, and the absence of such reference in the other.

There is an immense amount of good in the world: of mutual kindness, of parental affection, of conjugal love. This age, certainly in this country, is characterized by an active and general and efficient philanthropy. There seems to be a more profound and more operative belief than ever before manifested itself so widely, that it is an important part of the business of mankind to help mankind. The miseries and evils prevailing among us are searched into, their causes investigated, and earnest and able men are laboring to devise and to apply the best preventives or remedies. This is good, and very good. But the question still remains, what is the nature, what is the quality of this goodness? Is it natural, or is it spiritual? or, in other words, how far is its purpose limited to the improvement of human life, conduct, and condition in this world, and how much in it looks upon this life in its relation to another life?

I have been struck with a fact which I do not find noticed. There is now almost no form of vice, or misconduct, or misery, which has not its energetic assailants. Associations are formed to overcome it. The poet, orator, and essayist are all invoked against it, and answer to the call. But I have not been able to discover, in the constitutions, or platforms, or resolutions of these societies, in poem, speech, or writing, any argument for these reforms grounded upon the necessity of preparing by a life in this world for a life to come after this is finished. There may be some such; but there cannot be many, or I think I should have noticed them. And if it be said that some of these reformers are religious men, and refrain from bringing forward the religious aspect of the case from an apprehension that it might disturb united action, and so work a practical mischief,—what is this but an admission that the active and prevailing principles which are urging on these efforts to reform, must, in the present condition of human thought and character, be kept perfectly free from all reference to another life, or they would be obstructed and paralyzed? Does not this fact, with whatever qualification any may wish to give to it, help us to answer the question, whether this goodness is spiritual or natural?

But, if we look at the question on broader grounds, is there not a strong tendency in these days to regard conduct independently of motive, or at least so much of motive as refers to religious faith.

> "For creeds and forms let graceless bigots fight;
> He can't be wrong whose life is in the right."

This is entirely true, if *life* is here understood as comprehending *all* life,—the life of thought, belief, and motive as well as of act, — and if *right* is rightly understood. But is it not plain that these words are not, generally, so used? It is not meant that he who is kind, pure in conduct, benevolent and useful to all around him and injurious to none, is *a* good man (which he certainly is), whatever may be the reason or motive or faith on which this conduct is grounded; but that he is *the* good man, so good that it is only "graceless bigots" who ask that he should be still better, or think that he may be. Let it not be supposed that the *reality* of this goodness is denied or doubted by me, or its measure of efficiency — which may be spoken of more fully in other connections — in doing that work of preparation for another life, which is the object of this life, although *not* the object of that goodness; nor let it be supposed that between the natural faculties, intellectual or affectional, and the spiritual faculties, or between natural goodness and spiritual goodness, there is any antagonism. The truth is precisely the opposite of this. As the external universe, with all its relations, is perfectly adapted to the external needs of man, so is it perfectly adapted to the internal needs of man when it is called to the service of those needs. And as the natural faculties comprehend and utilize the external world for the purposes of this life, so the same faculties, when they are the instruments of the spiritual faculties and are subordinated to and animated by them, lose no jot of their energy or their success; and their work is thankfully accepted by the higher faculties as the foundation of *their* work.

Here is a good man. He is energetic and combative, and his benevolence turns his energy and combativeness against the mischiefs and miseries he sees around him. Or he is good in a more private way. His very nature makes it painful to him to see pain, and he does what he can to relieve or prevent it and give pleasure. But would this man abate his energy or grow less kind and benevolent, if he habitually looked upon all men as his brethren because all are the children of one Father; if he habitually regarded this life as a preparation for another, and all his efforts against human suffering were invigorated by the thought, that the great causes of this suffering are to be found in the moral depravity, or the moral faults, which mar the character and sicken the soul just where they must be clean and healthy if eternal happiness is sought for? Would the parent be less loving if he looked upon his children as immortals to whom he had been permitted to give life, and who had been placed in his care that he might train them for a happy immortality? Would the husband and the wife love each other less or with less purity and constancy, if they believed that their Father had given them to each other, as the very best assistants either could have, and as claiming from each other the best assistance either could give, in walking together in the path of unending life?

There is an antagonism between all that is natural and all that is spiritual, when that which is natural rebels against all that is spiritual, and if it had its way would suppress and destroy it. And there is another antagonism when the spiritual rises against the natural, and believes that its own growth and health demand the suppression of the natural. Asceticism thought so, and acted upon this belief; and this may have been well in times when the spiritual could not live unless the natural was suppressed. And they may well think so now, who not only feel the natural constantly rising against the spiritual, but find also in themselves nothing by which they can compel the natural to serve instead of ruling.

But all this is disorder. If it is good at all, it is so only because disorder makes it the best thing for the time and the person. There are no natural intellectual faculties which would not be helped and strengthened if they permitted the spiritual intellectual faculties to become their guides and their support. And there are no true spiritual faculties of the intellect, which would not guide wisely, and invigorate, and make more and more successful the natural faculties. And there are no natural qualities or affections, and no form of natural goodness, which would not find new strength, earnestness, and happiness, if they would accept spiritual motives and affections as their very soul.

V. OF SOME RELATIONS BETWEEN THE NATURAL AND THE SPIRITUAL FACULTIES.

The truth which answers all questions as to the relations between these two classes of faculties, or qualities, is that which tells us that this life is a preparation for the other life. But this truth is itself apprehended and perceived only by the spiritual faculties. In their present debility it is seen imperfectly, if at all. Probably multitudes would admit it in words; and would rather expect it to be said in any writing concerning man's nature and destiny. But everywhere in life, in common opinion, motive, interest, and conduct, we find overwhelming proof that this truth is seen but very dimly, and holds a very subordinate place and exerts but little influence in human thought. Hence the consequences of this truth must be seen but very dimly and imperfectly. But if this central truth in the philosophy ot human life and destiny were seen with any clearness, it would also be seen as a positive certainty, that this world must exist because of and in reference to another, and that the faculties which have this world for their proper object, must exist for the sake of those which have another world for their proper object. It would be seen and known that the lower should be the instruments of the higher, and would find in this instrumentality their greatest value; and that the higher need these lower as instruments without which their own work cannot be done as it should be done.

If we believe that this life is intended to be preparatory to another, we may believe that this world was intended to assist us in making this life a preparation for another. And that, for this purpose, this world is created such as to afford exercise not only for the natural faculties which would utilize it for itself, but for the spiritual faculties which would use it in this way of preparation. To some extent this has always been seen; and some religious men have labored to find in this evidence of the existence, the love and wisdom and providence, of God.

The men who are most energetic and successful in their investigation of physical nature may not know, or may forget, that this creation must be, in some way and measure, a record of its Maker; and a record, which, if we have not yet read it, can hardly be illegible in itself. If this be true, God must know it, and cannot forget it. And it may be that if the present condition of human nature makes it now impossible for these two uses of the external world, — for the knowledge of itself which the study of it yields, and the knowledge of Him which the study of it might yield, — if these two uses cannot now be brought into unity and conjunction and worked out together, — it may be that He can provide, and is providing, for them separately. If the scientists of to-day, in their devotion to external nature, are blind to all that is within it or above it, they may still be accumulating stores of natural science, which at a later age of the world and by other men will be made to yield their harvest of spiritual science.

Attempts of that kind have not been wanting in various ages. But it must be admitted that they have been feeble and ineffectual. If a poet has said that "an undevout astronomer is mad," he has not shown it to be so, nor is it more true of an astronomer than it is of any other scientist. We live in an age characterized by a marvellous activity and success of the natural faculties, and an equal torpor and debility of the spiritual faculties; but we may hope that another age will come, when the spiritual faculties will be roused and strengthened, and will do their proper work with an energy and a fruitfulness like those which mark the present activity of the natural faculties. Perhaps an early part of this work will be the laying hold of this external science, by those whose spiritual faculties are equal to the effort and can therefore make an appropriate use of all it has gathered for them, and show that the world which God made and pervades is a mirror of Him, and its activity of His activity. I indulge the hope that, when that good day — that brighter day — shall come, it will not be found needful to continue this division of labor with such an entire distinctness of work. But that the natural scientist will then be also the spiritual scientist; and that he will labor with not less earnestness and efficiency and enjoyment, when he is guided and animated by the experience that every day adds to his knowledge of the universe which God has made, and at the same time to his knowledge of its Maker. Then, at least, one long-lived delusion will have perished, — the delusion which tells us that this life and immortality are enemies, and that each must be guarded from the encroachment of the other. For it will be seen that life is but the beginning of immortality, and immortality but the completion of life; and that they have one Lord, and one Law.

There is one way in which the present predominance of the natural faculties over the spiritual faculties leads to a sad result, which cannot be understood without some consideration of a distinction between these two classes of faculties, which is seldom adverted to, and perhaps not often known. It is the distinction between the evidence which the natural intellectual faculties require and may find, and that evidence which the spiritual intellectual faculties require and may find.

The former ask for proof of a certain kind. All this proof must necessarily begin, as all proof must begin, with axioms, or truths not proved nor provable, but held to be true because they are seen to be true, and the natural reason declares that they must be true. But after these first and simple foundations are laid, a superstructure is built upon them, by gradual accretions of truth admitted upon a certain kind of proof. If mathematics enters into the inquiry, this proof must amount to what it considers demonstration. If the facts or laws of natural science are investigated, here too must be proof; not now mathematical, but resting upon facts which the senses may ascertain, and conclusions which may be tested by these facts, and which solve the problems of science; and only when they do this perfectly are the principles of science acknowledged as certainly true.

But there may also be progress in spiritual knowledge by the exercise of the spiritual faculties. This too must begin, as all knowledge and all proof must begin, with axioms; with truths conceded because they are seen by these faculties to be true. And these axioms can be seen to be true and certain only by the spiritual intellectual faculties. For the natural faculties have nothing to do with them; they can neither discover, nor recognize them, nor use them as the foundations for farther natural knowledge. And then when the spiritual faculties use these axioms, they use them in their own way. They do not build up on them a structure of belief every part of which is made certain by evidence or argument of like kind with those which are proper to natural science. There is nowhere what natural science would call proof. I will not now attempt to state the reason for and the good derived from this difference, but will only say that there never yet was, and there never will be, because there never can be, any religious truth which is not so given and does not so present itself to the mind, that it may be rejected if the mind be in a negative state in regard to it. Neither the central truth, that there is a God, nor any of the truths which cluster about this centre, admit of *proof* in the natural sense of that word. For spiritual truths rest on spiritual proof. They admit certainty, and certainty of a higher kind. They not only rest upon primary axioms as do natural truths; but in the whole structure of belief and faith, something of this character of axioms, of this acceptance of truth because it is seen to be truth and necessary truth, prevails. There is, in systematic religious truth, ample room for the most exact logic, the profoundest argument, the most cautious sagacity. But they are spiritual; they are true to their own work and their own character; they do not ask for and would not be satisfied with what natural science regards and requires as proof.

Hence we may see one difficulty under which scientific men labor with regard to spiritual truth. They have established habits of experiment and observation by the senses, and of rigorous ratiocination of a kind appropriate to the subjects of their inquiry. They owe their success — whether this means fame and position, or the pleasant consciousness of knowledge — to a careful adherence to this way of thinking and concluding. It is and must be very difficult for them to believe that there is or can be any truth which should not be sought and may not be learned in that way. But religious truth cannot be so learned, and should not be so sought for. This, however, is the only way they know of, for seeking after or learning truth. How can they escape the conclusion that religious truth lacks all basis, all confirmation, and is not truth but error?

With all the breaking down of religious doctrine, there is still much religious sentiment with many persons. It refuses to be suppressed. And as it demands some formal belief as its expression, it accepts that into which it was educated, or which it prefers among the many it may choose from. This religious sentiment imparts its own conviction to the belief thus held; and this belief thus becomes, in many cases, very strong. It is far better than unbelief; but it is not held on spiritual-rational grounds. This is so now; but it needs not be so, and will not always be so.

A distinction exists between the natural affections and those which are spiritual, analogous to that between the natural and the spiritual intellectual faculties. We have already said there is much natural goodness among men now; much that is real, active, and most beneficial. But in some minds it leads to the positive rejection of spiritual goodness. They do not desire, they cannot perhaps imagine, any higher goodness than they can see to be derivable from a due desire to make men live as they should live, and be what they should be, in this world.

They may or they may not express this rejection of all spiritual goodness. But it is there.

With many of those who are characterized by natural goodness, while there is no positive rejection, there is little thought of, or care for, or aim after, spiritual goodness. It is only nothing to them.

Let me try to illustrate this distinction or difference farther. And, first, as to the intellectual faculties.

As the external world is given us for two purposes, — one, to be a suitable home for us while we live in it; the other, to educate us for the other world while we live in this, — it is the lower or natural faculties which utilize this world as our home, and find pleasure in studying it and all its elements with no reference to any possible relation to another world; while the higher or spiritual faculties do or may lay hold of all the external truth which the lower faculties discover, and use it for their own higher purposes. Because man possesses these two distinct classes of faculties, every thing belonging to the external world may be seized and used by either class, in its own way, for its own purpose, and its own result.

Precisely the same thing is true of every thing of the inner world, of mind, thought, or affection.

Thus, there are certain ideas (I do not profess to use this word with metaphysical precision, if indeed there be such a thing) which belong to every rational mind; as the ideas of existence, of unity or plurality, of identity or difference, and very many others, and among them the idea of cause, or causation. I have nothing to say as to whether these ideas are innate or connate, or suggested, or original or derived, or constructed, all of which phrases and many more are used by those who treat of these subjects. I mean only that every rational human mind finds these ideas in it, when it has any occasion to make use of them.

For example, take the idea of causation. This belongs necessarily and inevitably to human thought. No man ever got rid of it, or lived a day or an hour without making use of it, however unconsciously. Natural common sense constantly accepts it, relies upon it, and makes use of it. But how do natural science and natural logic deal with it? They recognize only certain sequences of events: they say, we neither know nor can know any thing more than that, when a certain thing has occurred, another certain thing has followed, and this so uniformly that we believe it always will. This is true, say these faculties, as far as we know. And then they hold that we fall into the mistake of thinking that to be necessary which we first think only invariable. They suppose we know nothing of cause, or causation. And they are perfectly right, on their own ground. The natural powers of the mind do know nothing, and never can know any thing, about the origin or nature or reality of causation; although the idea of causation is universal and inevitable, and the natural faculties do themselves seize and hold it and use it in their earlier development, and continue so to do in their fullest development.

It is in every mind; and it is there for two purposes. First, that natural common sense, or the natural faculties in their healthy condition and normal action, may make use of it for all the needs and uses of common life or of natural science. But this idea is in the mind for another purpose also: it is, that the higher spiritual faculties may make their use of it; and that, so used, it may lift up the mind through the chain of causes to the Supreme Cause; and thus make some idea of God to exist in all minds but those incapable by degradation of thinking such thoughts, or those in whom the idea is crushed out forcibly by the natural faculties when they have become so perverted, so unhealthy, — insane, — as to aim at and succeed in paralyzing the spiritual faculties.

For the natural faculties may utterly pervert the

influence and effect of this idea of causation. They may come to the conclusion that the chain of causes does not extend upwards and end in a Supreme Cause; but that it is a revolving, a circular or endless chain. They may hold that the universe exists as a congeries of causes and effects, which produce each other; that it is sufficient in itself; that so much cause, or force, is in it, and operates in a definite and unchanging way in and through all the modes of action which, taken together, constitute the universe. They may call this universe God, or not; nor does it make any difference except in words. For Pantheism is simply Nontheism.

Then, the natural faculties deny that there is any such thing as causation. They can hardly help doing so, if they are true to their own logic. For all force, or potency, or causation, is from God, and is Him in action. It is the putting forth of His infinite power which sustains all being by continual and constant action; and all the laws of being are but the methods of that action. He is the First Cause; and all subsequent causation is the extension of His primary causation. This, the merely natural faculties cannot perceive or understand or acknowledge; because the whole subject lies beyond their scope.

Now let us turn to the natural affections, and suppose them carried to the utmost point of philanthropy, private or public. Men ask what more can they do? They are right in this question, on their own ground. They not only stand on earth, but they do not lift their eyes from earth. They regard only earth and earthly life. And, for this life, what more can be done than natural philanthropy can do? But this philanthropy takes no cognizance of the truth that whatever we can do to help a man through an unending life is of far more worth than all we can do to help him during this transitory life. Natural philanthropy does not know, cannot possibly know, that it would itself be far more ardent, far wiser, far more successful in its own work, if it were the expression, the instrument and embodiment of a spiritual philanthropy.

VI. OF THE COMPARATIVE STRENGTH OF THE NATURAL AND THE SPIRITUAL FACULTIES.

As either the natural faculties or the spiritual faculties may have any measure of strength or of weakness, so the quality and character of a race, an age, an individual, depends mainly on the prevalence of the one or the other of these two classes of faculties, and also upon the degree of the prevalence of that which is the stronger. For either may be the master, but with only an uncertain ascendency; or it may have any degree of sovereignty, until it becomes despotic, and is fiercely hostile to the other, and seeks its destruction.

A general view of the condition of mankind, especially among the nations of Christendom, and most among those which may be considered as the most advanced in civilization, would lead to the conclusion, that at no period within the records of history have the natural faculties of men been so powerful, so active, or so successful, as in the present period. And that at no former time have the spiritual faculties, intellectual or affectional, been, in proportion to the natural faculties, so feeble, so inactive, or so unfruitful as they are now. If this be true, it is a truth which cannot be generally acknowledged. For it can be seen to be true only by those faculties which are now so feeble, and which, if not silent, speak in muffled and uncertain tones.

If this be so, the darkness which covereth the nations may be nevertheless, and we believe it is, that darkest hour which precedes the dawn. And we believe too that the day, which is about to come, already casts some light, a dim and broken light, upon the clouded sky.

These higher faculties certainly are not in all men, nor in most men, wholly silent. Perhaps in no man are they absolutely inert; for then it would be impossible to rank the man higher than the animal. They tell nearly all men that there is a God; that there is a future life; and that the life which follows death is in some way dependent upon the life before death. They tell these great truths very obscurely and imperfectly, and very variously; and everywhere their utterance is distorted by the influence of the natural faculties. These can, of themselves, tell us nothing whatever concerning God or immortality. When the higher faculties take some cognizance of these ideas, and at least hold them as topics for consideration, and the lower faculties claim the right to deal with them as their property, supposing themselves to be sovereign over all the domain of truth, — a claim now generally advanced by those who have cultivated these natural faculties most exclusively and most successfully, — these lower faculties can only deceive, can only darken the intellect. It is of the higher faculties that we must ask about these higher things. They will use the lower faculties, and all they teach, as their instruments; will find them most useful and wholly indispensable instruments. These two classes of faculties (and each of them is either intellectual or affectional), taken together, constitute the man. Let the two, the inner and the outer, co-operate, the lower in due subordination to the higher, and there is no limit, and there will be no end, to their progress. There will be error in their utterance, but it will be that which is due to the mists of morning and the shadow of the earth; and it will gradually disappear as the light grows stronger. If there be danger in listening to their voice, we shall be effectually guarded against it if we listen reverently and humbly, seeking to hear in their voice the voice of God, and believing that if what we hear is true, it is then His truth, given by Him to men, through the faculties which He has given to men that through them they may hear Him and know Him. And the hope is not presumptuous, that in this way we may reach some elementary knowledge of the relations between the Infinite and the Finite, between God and Man.

It is perhaps impossible that either the natural or the spiritual faculties should be utterly extinct or wholly silent. It may never be so in any age, or any race, or any man. But one of them may be far more powerful than the other; for upon either of them the whole strength and force of the race or the man may be concentrated, leaving the other exceedingly feeble. In the early ages of civilization it would seem that religion exerted great power. Of its character or quality I say nothing, but only that it had great power. One test of this may be found in the remains of their architecture. These are all, or nearly all, religious structures; many of them of enormous magnitude, indicating that the whole strength of the nation must have been employed in their erection. And such remains as we have of their literature or philosophy lead to the conclusion that their religion — such as it was — mingled with all their thoughts, giving to them a certain tone and color.

Most truè it is that religion was then very largely a superstition, often irrational, and sometimes cruel. But it was still religion, however false, mistaken, and perverted. We must not forget that the question of the comparative vitality and force of either the natural or the spiritual faculties is the only one we are considering now, and is perfectly independent of the question whether the spiritual faculties, intellectual or affectional, were exerted wisely or foolishly, usefully or mischievously. The spiritual faculties may be very feeble, and yet in all they do be wild and erroneous. They may have any measure of strength and predominance, and be equally wild and mistaken.

This superiority in force and activity of the spiritual faculties comes down to historic ages, and in Greece fills the poems of Homer with Gods and Goddesses; and holds such a place in the mind of Plato, that one of his critics, perhaps the ablest, has said that it seemed impossible for him to pursue any train of thought without beginning and ending with the idea of God; and it built the Parthenon. The latest historians of Rome attribute the invincible strength of Roman character in their earlier ages to the force of the religious sentiment and the power of the sanctions of religion, derived, as some of them say, from their Oscan ancestry. And if we smile with equal sadness and contempt at the deification of Augustus and his immediate successor, let us remember that there was a reality in this to many minds. Nor is it a wholly meaningless fact that among the offices by which Augustus secured that aggregation of all political power which made him a despot, was that of Pontifex Maximus, — High Priest.

If we come to the Middle Ages in Europe we may remember the Crusades. Then too were built those cathedrals which remain the unequalled wonders of the world. We call those centuries "the dark ages," because there was so little of what we call light. We may well call them so if we look only at the fantastic follies which seemed to darken human thought on every subject. We may say it was a mere folly which threw Europe upon Asia in the Crusades; but it was a religious folly. And let us not forget, if we are seeking to know only whether religion was strong or feeble in those days (not whether it were wise or foolish), the diversified but irresistible evidence that it entered, in some form or other, and with some result or other, into all speculation and nearly all action.

If we come down to our own times, the field is too broad to be examined in any detail. If we suggest that the temple of Luxor, covering some thousands of acres, and proving by its stupendous fragments the enormous amount of labor it must have cost, — if we suggest that this one temple in the estimation of skilful engineers consumed as much of labor as the iron road which is now a pathway from the Atlantic to the Pacific, or the canal of Suez which now connects Europe with Asia, we do but touch upon a course of inquiry which would lead to the conclusion we have already expressed concerning the character of this age in comparison with that of its predecessors. Whether it were wiser and better to make that long railroad rather than a temple as large as that of Luxor, is not now the question; which is only this: was it the spiritual or the natural element of the human character which built the temple or the railroad?

But, we repeat, these spiritual faculties are never utterly extinct, or wholly silent. Man must have some idea, some thought of God, some thought of spiritual things, even if it does not amount to belief, or even if it amounts to denial; or he would cease to be man. And these spiritual faculties are far from being extinct or wholly silent now.

VII. OF THE IDEA OF GOD.

What is our idea of God; what idea of Him, the Infinite, is possible to man?

To answer these questions we must revert to the fact already fully stated, that man possesses at his birth two natures, meaning by this word "nature" the complex of all the qualities, faculties, and tendencies which he has at birth. All of these are then wholly undeveloped; and all are capable of indefinite development.

All of them which taken together constitute one of those natures, I have called the natural faculties; they may be said to constitute the earthly nature. All of them which constitute the other and the higher nature, I have called the spiritual faculties; they may be said to constitute the heavenly nature.

The earthly nature is given for two purposes; one is, to possess and comprehend and utilize the earth and all that belongs to it; the other is to subserve the purposes of the heavenly nature. For this nature is given because man is immortal, and it is such that, by a due exercise and cultivation of it, man's immortality may be happy; that is, heavenly. And the earthly nature is in all respects and particulars adapted to become the instrument of the heavenly nature, and the means by which man, while living on earth, becomes prepared for heaven.

The next question I would consider is, in what way and by what means are these two natures, first, or in the beginning, exercised or developed? Or, what is the foundation of their development?

As to the lower or earthly nature there can be no uncertainty. Sensation is the foundation of all its activity and the beginning of all its growth. An infant, a child, or a man, would be perfectly devoid of thought, consciousness, or action, if he never had sensation. He learns first from his senses, and afterwards from his thoughts about his sensations. In the beginning what he learns from his senses is obscure and imperfect, and his thoughts about his sensations even more so. As growth proceeds, what he thus learns, and what he thinks about what he thus learns, becomes more distinct. This improvement goes on indefinitely. The senses never tell him every thing, and what they tell him often needs correction. But his mind makes this correction, step by step, and gradually enlarges and rectifies his knowledge of the universe about him, beyond what the senses alone can ever teach him. This his natural faculties do by rationally using instruction derived from his senses; for that instruction remains for ever the foundation of all he learns or can learn about his earthly home.

If, now, we ask what is the foundation of the earliest exercise and the continued growth of the higher, the spiritual, the heavenly nature, the answer is this: what sensation is to the lower, the natural or earthly faculties, Revelation is to the higher, the spiritual or heavenly faculties.

This is not so obvious. Or rather this truth can itself be apprehended only by the higher or spiritual faculties. This age is characterized by the comparative debility of these faculties, and none escape the influence of their age. It must then be difficult for the spiritual faculties to see clearly their own foundation, while the lower faculties do see clearly their foundation. But were the spiritual faculties as strong and free and unimpeded in their perceptions and in their work as the natural faculties are, they would see clearly that they rest wholly upon Revelation. For they would see that it is simply impossible for the natural faculties, however powerful or however cultivated, to form, of and from themselves, the first or simplest idea of any thing more than earth and earthly knowledge and earthly life.

They could not have or give the least idea or suggestion of any life after death, or of an invisible, personal, infinite God. For all this, man must depend on Revelation. I mean by this word Revelation, truth taught by the Lord directly to mankind in such places, in such times, in such measures or forms, and by or through such agencies, as He sees to be best. I shall recur to this subject and endeavor to say something of these various forms and agencies. Now I would say only that I consider the primary and constant law to be this. All truths or germs or elements of knowledge, which relate to this world or life therein, must come to the mind from this world, through sensation; and are then committed to the lower or natural faculties, which are adapted to these truths and to this world. Precisely so all truths or germs or elements of knowledge which relate to the other world and life therein, must come to the mind from that world through Revelation, and are then committed to the higher or spiritual faculties which are adapted to those truths and to that world.

The spiritual faculties, receiving thus the idea of God from Revelation, impart it to the natural faculties. The spiritual faculties can receive it at first only in its simplest form, and they can impart it to the natural faculties only in such wise and measure as they are ready to receive it. In such wise they do receive it, and may thereafter bring their resources to the aid of the spiritual faculties. Thenceforward, as far as these two classes of faculties are free from perversion and abuse, and work together in harmony, the lower subordinated to the higher and the higher instructing the lower,—just so far the whole man grows in intelligence and wisdom, and the ideas of God, and of immortality, and all the ideas which belong to religious truth, grow in force, distinctness, and development.

What, then, is the first and simplest form of the idea of God; its seed-form so to speak, out of which may grow all possible development and enlargement?

The first answer and a most important answer to this question is, that this idea must, while given by Revelation, rest upon and be clothed with ideas derived by man from himself. All knowledge originates in sensation, and all thought must begin with the objects of sense. But in man it soon rises into consciousness. And if we ask ourselves what we know, we become aware that we know our own existence more certainly than we know any thing which the senses teach us, or which we can learn from thought employed about what the senses teach. We cannot doubt our existence because that doubt proves existence. And we learn our existence from consciousness.

As soon as consciousness passes beyond this first and simplest idea of personal existence, it begins to ask itself what it knows of our personal individuality. It knows this: that we have affections, or purposes, or a motive force within us; that we have thoughts; and that we act because our affections through our thoughts lead to action, and we have power to act. Then the spiritual faculties may begin to impart what they have learned from Revelation. And then and thus men acquire the idea of One, in whom affection, thought, and power exist in a far higher degree than in ourselves; and we go forward until we think that they exist in God without limit and without qualification. And then, as our idea of God enlarges, we attribute to this One the creation and the government of the universe. They who now with most success are investigating the earliest forms of human thought and belief find much indication that, in ages so remote that but faint intimations of their characteristics have come down to us, this first and simplest idea of God prevailed.

It is very common for writers on these subjects to object strongly to all "Anthropomorphism," as they call all effort to construct the idea of God out of our idea of man. I have no fear of this word. It is a wrongful anthropomorphism which would degrade and lessen God into likeness with ourselves. It is a rightful anthropomorphism which seeks most earnestly to rise into a likeness to God. If He were altogether other than we are and in all respects that which we are not, it would be impossible for us to form any idea of Him, or have any knowledge of Him, or any faith in Him, or any affection for Him. And this is precisely the result to which some are led by their dread of likening God to man. It is because He has made us in His image and likeness, that every religious man — of every name, or age, or faith, consciously or unconsciously — regards the growth in resemblance to his Father as the measure of all progress and the goal of all good life.

They who thus fear and dislike anthropomorphism are wrong in so far that it is impossible to construct the idea of God out of any other elements than those which consciousness gives us. But they are right in denouncing it as open to great danger. It is the spiritual faculties given us for that purpose, and taught and aided by Revelation, which so construct, out of the elements given us primarily by consciousness, this idea of God. It is only the spiritual faculties which can give us this idea. But then the natural faculties receive it, and afterwards will be heard about this idea; and they will do their own work upon it. And unless they acknowledge their absolute subordination to the higher faculties and bear a true allegiance to them, they can say no word on these subjects that is not false, and can do no work upon this idea of God that is not mischievous. Therefore it was that the natural faculties, qualities, and proclivities of men laid hold of this idea, and burdened and distorted it with fantasies and idolatries of every description. But the spiritual part of man neither slumbered nor died. It worked, however impeded and fettered, for the salvation of men. It could not cast off the monstrous lies and follies which the natural part of man devised and loved. But even in them it did what good it could. And far better was it and is it for man to worship all the Gods or idols of superstition, or the sun and stars, or stocks and stones, than to have no God, and no sense of worship. For one who passes into the other world with no sense of worship, is to the last degree unprepared for that heaven where life is worship; and worship is the expression of love to God; and the love of the neighbor is born from love to God, and manifests itself in constant usefulness.

The Bible tells us that God made man in His own image and after His own likeness. This is much worse than nothing to those who deny the sanctity and authority of the Bible. They call it a falsehood springing from the tendency of the human mind to liken God to man. And because they deem it a falsehood they argue from it against the authority of the Bible. But is this tendency always and altogether mistaken? One answer is, that it is inevitable; it is of the very essence of our nature. Other answers will be given as we go on.

VIII. OF THE GROWTH OF THE IDEA OF GOD.

We have seen what constitutes the primary and the simplest idea of God. Let us try to see what more the spiritual faculties, using the instruction of Revelation, and unimpeded by the natural faculties, and helped by them so far as that may be, can do for this idea.

We shall see that all there is in man that is not material belongs either to what he loves or desires, or else to what he thinks and knows; and we may give to the whole complex of all affections, desires, or motives, the name of Will; and we may give to the whole complex of all thought and intellectual process or conditions the name of Understanding.

We have seen that only by exalting them to the utmost, and taking from them limit or qualification, do we form, by our spiritual faculties, an idea of

God. The simplest name to give to this first and simplest idea, is that of a Divine Man. But the spiritual faculties, by which we construct this idea, then using the instruction of Revelation, help us to protect and preserve it from the errors and fallacies which the natural faculties would cast upon it. And holding it in its purity, they go on to give to this idea distinctness and extension; and they develop such further ideas as grow out of this idea, under their influence.

They begin with guarding this idea from the fallacies of sense and natural faculty. They assert a Divine Man. He is their God. But they deny to Him shape, or any of the limitations without which the natural faculties cannot take one step in any direction.

They deny to Him shape, or local form or presence. But, if they do this, how can they retain any idea whatever of God as a Divine Man? The natural faculties certainly cannot; but can the spiritual faculties?

In the first place, let it be said, and remembered, that the spiritual faculties, unless grievously perverted, are humble. They know their own feebleness, their own immaturity, their own limitations. They know the infinitude of truth. They know that it must come to us and be seen by us only gradually;

and that, come as it may to any beings anywhere, it must so come, that, when its infinitude is remembered, it is seen to come only little by little. They know that, as there is no limit to the possible progress of truth, so there is no limit to the possible advance of their capacity to see and comprehend the truth, however feeble the beginning. They know that it would be the worst of follies to cast off or doubt a truth they do see, because around it lie difficulties or present impossibilities, which do not throw them backwards, but only point out the forward path which their future progress is to pursue.

IX. GOD AN INFINITE PERSON.

An Infinite Person. The difficulty lies in reconciling these two words, or the ideas they express. But the higher faculties look to the higher things in man. They look to his affections and his thoughts; for these constitute the man. The body is adventitious; it is his instrument, not himself. Changing its substance every day, and indeed with every breath, it is no essential part of the abiding individual man. It is not of the essence of the Person. We can know each other only in the body, because that is the covering and instrument of our senses. But what we know, unless our knowledge is very superficial, is the man within the body. Our habit of looking only at and thinking much of the body, and of getting access to the person only through the body, makes it difficult for us to think of God as man, and yet not think of Him as in a body as we are. And while the higher faculties are controlled or importantly affected by the lower faculties, it is impossible to do this; and hence the old mythologies and idolatries. But these higher faculties, in the degree in which they are liberated from the influence of the lower faculties, see clearly that this difficulty belongs to them and not to the object of thought. And being sure that He is a Divine Person, and thankful for this certainty, they wait with hope and patience for the gradual increase of their capacity to conceive of Him adequately: sure also that, while this increase will be constant and eternal, it will never enable the finite fully to comprehend the Infinite, as it is in Itself.

The higher faculties are employed upon affections and thoughts; and these are beyond the scope of Space and Time. And while they acknowledge the difficulty of thinking without reference to Space and Time, they know that a reference to them of things or thoughts which do not belong to them causes error, and against this error they are on their guard.

God is a Divine Person; a Divine Man; consisting of Love and Wisdom, both without limit, and this Love is infinitely active and productive through this Wisdom. Here then are infinite love, infinite wisdom, and infinite activity or power. Taking with us this idea of God, let us now ask what are the relations between this God and Man. Nor can we answer this question except in the degree in which we can account for and explain Existence and the Laws of Existence.

God is a Person, of or possessing infinite Love. What is Love? What it is in its perfection and infinitude we cannot completely know. Nor can we know any thing about it except by that analogy to man to which we are compelled to refer continually. Of what love is in man we know something. We know this. That love desires to do good to its object, and make its object happy. It is in its fulness in God. It must be in Him an infinite desire to make others happy. It must be in Him an infinite desire to have, and that He may have, to create, those whom He may make happy. This, then, is the final cause of creation. And it must also be the final cause, or the constant rule and guide, of this Love in the government of the world which it has created. Here then we take our stand upon a simple truth which must be appreciable to all; and which has in it nothing of novelty, for the higher faculties have always suggested it; and it presents nothing of difficulty to those faculties, although meaningless and mere nothing to the lower faculties. This truth is, that God creates and governs the universe, that He may satisfy His own Infinite Love.

Let us take one step farther. Love desires to make its objects happy. In its fulness it desires to give all it can, to give itself, all it is and has, to the beloved object. Love is in its fulness in God. He then desires to give all He can, Himself and all He is and has, to the objects of His Love. He must desire to create them such that He may impart Himself to them. Here then we have reached a truth which will guide us and may help us in all inquiries as to the purposes and the methods of divine creation and government. It is not so simple as the former truth. But we believe it to be intelligible; and that it will be seen to be true by those to whom it commends itself.

This truth, stated as well as I can, is, that God creates and governs the universe, that He may have objects of His love, to whom He may, in the degree in which it is possible, impart Himself and His own happiness.

It may be well to remember that if God be Love,

or if He have love, He must desire to give happiness to those whom He loves. But His own happiness must be full and perfect. We cannot think otherwise. There can then be no happiness more than his, or other than his; for then it would be wanting to or absent from a happiness which being perfect can have no want. He must then desire to impart of His own happiness to the objects of His love. Now what is His happiness; what does it consist of; what does it flow from? Here, as elsewhere, in thinking of the Infinite Divine, we may exercise the powers given us, upon the truths given us, that we may know Him; and, feeble and immature as these powers may be, they may give us at least some elementary knowledge of Him, which, however little it is, may yet be sound and just. To the question, What is the happiness of God? we answer, that it must be that of Love, exerted through Wisdom, in Effect; that is, in use or work, which work is the creation and support of the universe. Then it would follow, necessarily, that this is the happiness He must desire to give to men. And this truth will be commended to us, when we see that it offers an explanation of the constitution of man and of the universe which is his home. We find that man, forgetting for the moment his body, consists of Will and Understanding. And why?

Because God creates in him a Will, as a recipient of His own Love; and creates it such a recipient, that it may receive into itself Divine Love, and that the Love so received may become all there is in man of affection, desire, or motive power. He also creates in man an Understanding as a recipient of His own Wisdom, and such a recipient that this Wisdom becomes therein all there is in man of thought, perception, or reason. But this is not all. The Divine happiness does not consist merely in the possession of Love and Wisdom, but in the activity and productiveness of the Love through the Wisdom. This, too, is infinite and constant; and the universe is its continual Effect. In whatsoever is, from the smallest grain or cell, and far below this, from the primary atoms or elements of matter, up through suns and earths and systems, everywhere the Love of God, in His Wisdom, is ever creating, producing, governing, sustaining. This is His infinite activity, and from it His happiness. He desires to give of this to man. And therefore He clothes man's will and understanding with a body, as He clothes His own Love and Wisdom with the universe. He makes man's body such (a natural body here, a spiritual body hereafter) that all there is in him of affection or desire may, through his thought, act, produce, and be of use. And all this use or work of man, miserably feeble almost to nothingness as it is, wherever it has any existence whatever, and so far as it is not polluted by selfishness which is the opposite of love, is in the image and the likeness of the infinite usefulness of God.

X. MAN IS IMMORTAL.

This explanation of the constitution of man may suggest many questions. One of them is this: How can we suppose that the very, very feeble reception by man of the divine love, wisdom, and usefulness, and happiness thence, can satisfy the infinite desire and purpose of God? We cannot, unless we suppose, farther, that man is immortal; that he lives through eternity; and that he is made capable of growing in love, in wisdom, and in usefulness and happiness, through eternity.

But, limiting our thoughts at present to this life, another question may be this: If all of man's love, wisdom, and usefulness are God's love, wisdom, and usefulness in him, where does his hatred, his folly, his selfishness, his wrong-doing, come from? How does love, proceeding from God, become in man hatred; how does divine wisdom become human folly, and divine usefulness human mischief?

The answer to this question must reveal the very heart of all that we can know of the relations between God and man.

That answer is, that God gives life to man, by giving His own life to man, to become in man MAN'S OWN LIFE, his own, his selfhood.

This is a truth very liable not to be understood at all, and to seem wholly unintelligible, or to be misunderstood. It is a truth, but only one of two truths which together make a whole. The other truth is, that the whole of man's life is given by God, of His own life, instantly, incessantly, always.

The whole truth is new on earth. I do not mean that I have discovered it, for I have learned it. But it has been delayed until the human mind was in some readiness to receive it. It is now new in human thought, and that readiness is most imperfect.

The whole truth, stated as one, may have this form. Man lives because there is a continual flow of God's life into him to become man's own life.

My endeavor now will be to explain this truth; for, being understood, it is nothing less than the key to the mysteries which belong to the relation of God to man and of man to God. I undertake this with the certainty that I shall do it imperfectly, and with the fear that I may fail to place this truth within the reach of those who would welcome it. But I will do what I can.

First, then, as to the first element of this truth; namely, that the life of God flows into man continuously and always. He did not impart life to the father of the race, and give to that life the power of independent continuance through all the generations of his offspring. He did not give life to each one of us when we began personally to be, and such a life as would abide for ever, or even until our death, without farther gift of life. But at the first moment of each one's existence He gave the life of that moment; and, at every succeeding moment either in this world or the other, He gives the life of that moment. I am obliged thus to speak of succession, and continuity, and moments. And yet the spiritual faculties, by which alone we can form or approach an idea of the Infinite, must be feebly exercised if they do not assure us that with God there is no Time. That belongs to us. Eternity is something more than a mere indefinite aggregation of time. What it is we cannot adequately know, because while we are here our minds cannot wholly escape the controlling influence of Time and Space. Neither can we adequately conceive of a personal existence without

Time. But we can detect and avoid the errors which come into our mind when we connect the ideas of Space and Time with that which we must know to be independent of them. When we say that God exists independently of time, we may not be able to conceive adequately how this is, or can be. But we may be able to detect and avoid the error and the obstruction to thought caused by the influence of this idea of Time when we connect it with the idea of God.

It is one of those errors to impute a past or future to God, as He is in Himself. What He did, He does. He did not give, and stop giving. He gave and is giving. We may formulate this by saying that existence is perpetual subsistence. We subsist from God, and from His gift of Being to us. We exist continuously because we continue so to subsist from Him.

This is a hard doctrine. What is more common than the notion that every man has at birth a certain measure of vitality, and, if his life be not cut short by disease or casualty, he will live on through the periods of growth and decay, and die, sooner or later, when that vitality, be it less or more, is exhausted? There may be so much of truth in this as to the body, as can be found in the fact that this body is weaker or stronger as inheritance or other circumstances may determine, and will continue for a longer or a shorter time to be the clothing and instrument of the man. But Time, the word and the thought, belong only to the body and its world, and to the faculties employed upon them. They have no reference whatever to the man himself, or to his soul. This will not die. But from the moment it begins to live through its unending life, it lives by instant and incessant communication of life from the only Source of life. Every affection or feeling or emotion, every thought and whatever pertains to thought, comes to him, then, when it exists. It comes to him as life, modified by the instruments through which it passes and the method of transmission, and is thus exquisitely adapted to become and constitute his life.

Then, when I recur to the second of the two propositions which constitute the truth I have stated above, — when I say that this divine life is given to man to become and be his own, perfectly his own, his very self, I state a still harder doctrine.

Because man's life coming to him from God is given to him to be his own, he has power over himself, a power of self-determination, which, as to all spiritual things, and all that belongs to his spiritual character and destiny, is complete and perfect.

He has freedom, *because* his life is his own. This

freedom no more originates in him than does his life. It is his, because it is given to him. It is his, because it is given to him with life; and it is given to him with life, because it is an element of life. It is his, because life is given to him in such wise that he cannot but be free; he must be free for the very reason that his life is given to him to be absolutely his own, and in that life is freedom.

XI. OF FREEDOM.

The idea of freedom is one of the primitive ideas of consciousness, and is in itself so simple that no one doubts what freedom is until he begins to confuse and obscure his thoughts by the effort to make that plainer which is in itself perfectly plain. We define any thing, or describe it, by referring to plainer and simpler thoughts or truths, and asking them to cast their light over the thought or thing to be defined. But freedom cannot be thus defined, for the very reason that it is itself the plainest and simplest thought. We know what freedom is, also, by the checks and limitations to it, as we know what light is by darkness. In physical matters these limitations are constant and universal. While every man has something of physical freedom, no man has all. So it is with natural freedom, which is above mere physical freedom, and distinguished from it by including all the freedom we have in the exercise of all our natural faculties, whether intellectual or affectional.

Here, too, freedom is always limited and qualified. The reason is, that as temporary life on earth is given as a means of preparing for the unending life that comes afterwards, so all our natural faculties are given us to be used as the instruments of our spiritual faculties in this work; and all our merely physical and natural faculties are restrained in their use and exercise so far as may be required by the *end* for which all are given. This end is spiritual preparation, and it is placed in man's power. Not one particle can he do of it or for it, but by the exercise of the spiritual life and power given him for this end. But this work can be done as it is intended that it should be done, and as the highest happiness possible for us requires that it should be done, only by man's co-operation, and only when man in the exercise of his spiritual freedom does his part of the work. Therefore is his spiritual freedom free from the limitations which belong to his natural and physical faculties. And only by the spiritual faculties can spiritual freedom be judged of, or made use of, or even recognized as existing.

Very few topics have been the subject of more

earnest reasoning and speculation than that of human freedom; but to very little purpose. Only the lower, only the natural faculties have been, for the most part, brought into action in relation to this subject. And the body has but little freedom. Man goes through this life, as to all his lower nature, in fetters. This is true of all the natural faculties whose proper scope is the body and its home. And, while bound and shackled by these fetters, religious men have entered into these questions, and some of them have founded religious doctrines upon the views which their natural faculties led them to take of spiritual freedom. For it is just as possible to bring the natural faculties to bear upon the spiritual topics suggested by the spiritual faculties, as on any other. Unless they are subordinated to and directed by the spiritual faculties, they do not change their nature, nor their method of working, nor the character of their results, when they are employed upon topics higher than themselves. They have no aptitude for these topics, and can lead only to fallacy and falsity when applied to them. I wish it were possible for me to present this truth with the clearness which belongs to it. The higher and the lower faculties coexist in every man. It may be that there are those in whom the higher faculties are almost wholly suppressed, but such cases must be rare.

It may also be that there are those in whom the higher faculties refuse all fellowship with the lower and suppress them almost utterly, but such cases also must be rare. The lower faculties can of themselves, as has been repeatedly said, teach nothing, know nothing, think nothing about God, or the Infinite. The higher faculties can employ themselves to the best purpose for the objects they may reach, when they make a rightful use of the natural faculties as their instruments; but never when they permit these lower faculties to bring what is their light as to natural things, but their darkness as to spiritual things, to cloud the spiritual faculties in their investigation of spiritual things, and distort and falsify their conclusions. And yet this has been the commonest thing in the history of thought and belief. Men of the greatest ability have devoted themselves to religious inquiries, but have brought their natural faculties to mingle in these inquiries, and perhaps to exercise a controlling influence therein. A reason for this has sometimes been their consciousness of the unusual vigor of those faculties and a pride in this superiority, and when in their religious inquiries they found themselves pushing questions into the nature of the infinite, farther than the spiritual faculties were able to answer them, they resorted to the natural faculties to work out for them some scheme or theory which should at least seem to answer; and they did this unconsciously, because they did not recognize any distinction between these faculties. An answer to such questions given by the natural faculties must necessarily be mistaken. It may well be that the spiritual faculties cannot now give a complete answer, and, conscious of their present limitations, make no attempt to do this. But the natural faculties have no such hesitation. In their own way they construct an answer, which, because it is founded upon reasoning which is only true in its application to the finite, necessarily misleads when it is applied to the Infinite.

Let me take an illustration from this matter of freedom. Calvin, by his method of reasoning, came to the conclusion that it was inconsistent with the being and government of God that man should have spiritual freedom. He was right and unanswerable on his own grounds. Man's freedom was inconsistent with the being and government of Calvin's God. Because the idea of God suggested by the spiritual faculties was surrendered to the natural faculties, and therefore his God was one whose being and character were measured and defined by the natural faculties. The perfect proof of this is to be found in his doctrine of *pre*destination, or election and *fore*ordination. For this idea, and the whole argument by which it is supported, become nothing, when we refuse to attribute man's *time* to God. For if we refuse to attribute time to God, we shall no longer speak of any thing in Him as *before* or *after*. Very little reasoning suffices to satisfy us that *time* cannot be an entity; that it is but a method of perception, or result of intuition, or a law of thought. What, then, must be the result of reasoning founded upon making that which belongs to man only, and his way of viewing things, a positive fact in itself and of the very essence of the infinite God?

So too with Jonathan Edwards. Let not one jot of the credit due for extraordinary power of intellect be denied him. But the intellect has its natural powers as well as its spiritual powers; and when Edwards reasoned, whatever might be his topics, he exerted the natural powers which in him had great vigor, and he did not subordinate them to his spiritual faculties. Both Calvin and he, after asserting and maintaining with all their might predestination and the absence of human freedom, then labored strenuously to reconcile this fallacy with human duty and responsibility. But they labored wholly in vain. Words may assert any thing. A man may say that he believes black to be white; but he does not believe this, for the reason that it is simply impossible to think so. If a man really believes that

Omnipotence decreed *before* he was born that he should go to heaven or to hell, he cannot believe that by any act or effort of his own he can change his destiny.

And now it may be asked whether when we say that the spiritual faculties assure us that time, or its incidents, effects, or conclusions, can have nothing to do with the Eternal and Infinite, we say also that these faculties can teach us *how* the Eternal thinks, knows, and acts without time. Certainly not. What they can do is to protect us from the error derived from supposing that He thinks or knows in time, and under the conditions of time. They do not reconcile the infinite knowledge of God with man's freedom; but they show us that no reconciliation is required, because it is only our lower and natural faculties which place any antagonism between His knowledge, or between knowledge as it must be in Him, and the freedom of man. They show us that there is no antagonism except between our finite mode of knowledge indefinitely enlarged and human freedom; and that His mode of knowledge is not our own indefinitely enlarged, for it must be other than our own, and precisely other in that particular from which alone comes that antagonism. We cannot think as He thinks, for we are finite and He is infinite. But well may we keep our faith and trust in the perfect wisdom which *foresees* all our needs and perils, and provides for them. It is right that we think of Him and speak of Him thus, for this is the form which His infinite care for His creatures puts on when it comes down within those bonds of space and time which belong to us. But let us not lay these bonds on Him, nor argue from them as if they belonged to His own essential infinitude.

The spiritual faculties declare that they cannot comprehend the mode and manner of infinite thought, and therefore they cannot investigate the compatibility of that thought with human freedom. They are certain of both of these facts. They are certain of freedom from consciousness; and this consciousness is confirmed and illustrated by ample and powerful reasons and considerations drawn from what these faculties perceive or learn concerning the relation of God to man and of this life to the next. They are certain that God is infinite, and, as He must possess knowledge, that His knowledge must be infinite. Being certain of these two facts, they are certain that in some way they are compatible and reconcilable; but that they cannot be clearly and fully seen by us to be so, for the simple reason that our finite faculties cannot have clear and full and adequate perception of the Infinite. They are certain of both facts, as certain of the one as of the other, and therefore certain of their coexistence, although this may be to a large extent incomprehensible by us in our present condition. And then they are equally certain that all denial or doubt of either of these facts, and all disturbance of mind from the difficulty of reconciling them, must be caused by the efforts of the natural faculties to take within their inquiry and determination questions beyond their reach and scope, and above their proper functions.

Not only the religious men I have mentioned and many others have dealt with this question on natural grounds, but philosophers have gone deeply into it. Some of them have come to the conclusion that there could not be in the nature of things any such thing as freedom. The main argument which has led to this conclusion is, as has been already intimated, that every effect must have its cause; that whatever is, is the effect of its cause; and that that cause could not but produce that effect, and could not produce any other effect. Hence it follows that whatever we feel, or think, or do, had its cause, and that cause its cause, and so through this chain of causes and effects the result we know came by inevitable sequence. So they have concluded. And to what purpose? None whatever.

They never produced on themselves or others much, if any, actual effect. No man ever lived a day or an hour without exercising some freedom, and knowing that he exercised it; for every man is more sure, infinitely more sure, that he has freedom, than he can be of any of the premises on which any such reasoning rests. Whence comes the fatal error of this reasoning? It has been already said, and is now repeated, that this error consists in utter ignorance of the distinctions between life and non-life. Of things without life the natural faculties may judge, and the reasoning above stated has much (though but a qualified) application to them. But, when we come to consider things having life, the first thing we must know is, that Freedom is an essential element of Life. There can be no freedom where there is no life; there can be no life where there is no freedom.

XII. WHENCE THE CONSCIOUSNESS OF FREEDOM?

Whence comes this consciousness, this certainty of freedom? In the first place, the consciousness of natural freedom belongs to the natural faculties. Then the consciousness of spiritual freedom belongs to the spiritual faculties. The spiritual faculties must not only be possessed by every man, but we hold that they must have *some* life, *some* exercise in him, or he would cease to be more or other than a mere animal. And man is not permitted to sink so low as this. Animals have and exercise their own measure of physical freedom; but they do not know it, because they cannot have the idea, or consciousness, of freedom. This they cannot have; but man can have it, and cannot but have it, because the consciousness of natural freedom belongs to the human natural faculties, and they *cannot* lose it. And the consciousness of spiritual freedom belongs to the spiritual faculties, and they cannot lose it. This is so, because if there were *no* freedom there would be no duty, and no responsibility; and without some sense of freedom there could be no sense of duty or responsibility. And no living man was ever absolutely denuded of this sense, even if he thought he was.

Without some sense of duty and responsibility, man could not take one step in advance from his natural condition. Human improvement would be impossible, and all preparation in this life for another would be impossible. The one end for which we begin life in this world would be perfectly defeated. That this may not be so, that the possibility at least of this improvement may always be open to all, this consciousness of freedom is given to all, and preserved in all as an essential and an abiding element in human nature. Hence, I repeat, it may be believed that no man anywhere ever lived and died, without some sense of duty and some sense of responsibility. It is too plain to require more than the mere statement already made, that, if there were absolutely no freedom whatever, there could be no duty whatever, and no responsibility. And it must be equally certain that, if there were no sense of freedom, there could be no sense of duty; for the idea or thought of duty implies as its necessary prerequisite some consciousness of freedom.

We may go farther. The sense of duty and responsibility not only requires the consciousness of freedom, but will be measured as to strength and quality and character by this consciousness. The natural faculties of man, in so far as they are higher than those of the mere animal, suggest to him the consciousness of freedom, and keep it alive. But they limit this consciousness within the bounds which belong to these faculties and their exercise; and these are the bounds of this life and all that belongs to it. They are very wide, they embrace all the relations and all the possibilities of this life; but there they stop. The sense of natural duty and responsibility is conterminous with this natural consciousness of freedom. It may be very wide in its scope and working; it may cover all the relations and possibilities of this life; it may have a word to say, and that a good word, whenever any question arises from any relation of any man to his neighbor in this life. It may command purity of conduct, benevolence, and honesty; it may tend to produce quiet and order and mutual assistance, and to secure and promote all the comforts of this life. But there this natural sense of duty stops: it goes no farther and it looks no farther. There have been instances — not many perhaps, but still instances — of men who displayed all these virtues, while they emphatically disclaimed any belief in, any reference to, any thought of, God, religion, or another life.

This is all that natural duty, founded upon natural consciousness of freedom, can do. But then the higher faculties, the spiritual powers of man, may awaken their consciousness of freedom and of duty; and they will deal with it very differently.

In the first place, to them it is a consciousness of spiritual freedom. They look upon this life as one of preparation and education for another. They look, and earnestly, upon this world, for the very reason that it is preparatory and instrumental to another world. In their sight, and in all their teaching and all their influence, the other life is primary. They recognize natural goodness, and gladly promote and assist it, even if they cannot raise it to spiritual goodness. For they see in it the means by which men may in some measure prepare for another life, although unconsciously, and with no wish or purpose of the kind on their own part. And this imperfect preparation is made possible for them, because they are made incapable of any better by their unwillingness to desire any better.

But when spiritual emotions and principles govern, or, to vary the phrase, when spiritual ends are sought, then justice, love, and purity are transfigured. They have lost no strength, no tenderness, no beauty; but all these are increased a thousandfold. The standard of all excellence is infinitely higher, and the motive to approach it infinitely stronger and more effectual.

Such is a spiritual view of freedom and a spiritual use of freedom. But it is a view which only the spiritual faculties can take; and it is a use which only the spiritual faculties, intellectual and affectional, can make. I have already referred to Professor Huxley; and I may well do so, for I can find nowhere a more prominent example of the natural scientist of this day. In an address recently delivered by him to the Young Men's Christian Association of Cambridge, England, he speaks of freedom thus: "I protest that if some great Power would agree to make me always think what is true and do what is right, on condition of being turned into a sort of clock and wound up every morning before I got out of bed, I should instantly close with the offer." Now these few words show plainly, first, that Professor Huxley is conscious of possessing Freedom, and is sure of it; and that he is also conscious that there is a right way and a wrong way of exercising this freedom, and is more or less pained by the consciousness that, possessing this freedom, he does not always exercise it aright. But we cite this passage mainly to say that it seems to us nothing less than marvellous that Professor Huxley does not see that by closing with this offer he becomes a clock and ceases to be a man; that he becomes a mechanism, and man is not a mechanism. Why does he not see that freedom is an element of human life, and so essential an element of human life that where it is wholly absent life is in no sense human? We can account for this marvel only by supposing that he cannot see this, because only the spiritual faculties can discern this truth. Spiritual or moral freedom is this constant and necessary element of human life for a certain purpose. Only the spiritual faculties can discern this purpose, only they can promote it; for it is the preparation of the man for eternal happiness by the rightful exercise of this freedom. Only the spiritual faculties can teach him or can help him to make this preparation; and they only can recognize the truth that freedom is an essential element of human life, because they only can recognize the purpose for which freedom is made to be this essential element.

Personally I may be doing great injustice to Professor Huxley. I speak only of what he says to the public. And I am glad to admit that he says to them some other things which cannot be reconciled with these.

There may be those who stand at either extreme; who are so wholly natural that nothing of spiritual life is in their thoughts, or who are so far spiritual that merely natural emotions or purposes have no power within them. But, if there are any such, they must be very few. In most men, if not in all men, these two classes of faculties or feelings mingle their influence. But it must be plain that these higher faculties will be strong or weak, predominant or subordinate, wise or mistaken, in proportion to the clearness of view with which the fundamental principles of all spiritual truth are seen, and to the force of the motives which are founded on these principles. And none of these lie nearer the foundation of all truth than those which assert the fact of freedom, and disclose its origin and purpose, and then throw their light upon the law of duty. To find these principles we must go back to the twofold doctrine which presents two truths, but forms of them one. Of these two, one is that human life is divine life given continuously and incessantly to man. The second is that this divine life becomes man's life by being given to man to be his own, and so make him to be himself.

These truths coalescing into one become the source of all we can know on this subject of life, freedom, duty. But they must both be held: they must be so united that the one suggests the other, and refuses to be seen apart from it. For, if either be held alone, it profits us little and exposes us to great peril; and if either predominates over the other, or eclipses the other, in the same measure will our thoughts on all these subjects be distorted or darkened. To believe that all our life is divine life in us, and to believe this alone, would deliver us up to many and gross errors. To believe that our life is our own, in the sense that it is self-derived and self-existent and independent of God, would lead us to more and if possible to still worse errors. But if we are able to see both of these truths clearly, and to see them as one truth, and then to see why man is so constituted, we are restored at once to a sense of the most perfect dependence upon God, and to an equal sense of freedom, of responsibility, and of duty.

It is more than easy for a man to believe that his life is his own. It is difficult, not to say impossible, to believe otherwise. This is indeed a universal belief, and it might almost be said a universal consciousness. Wherein then lies the importance or the novelty of this doctrine? It lies in its connection with the belief of the higher truth, that he lives by constant and perpetual reception of divine life which is constantly given to him.

It is natural to man to believe that he lives from himself; or if he supposes that his life came to him from his parents, then to believe that human life originated in and from itself; or if some religious influence leads him to refer the beginning of human existence to God, then to believe that it was given once for all, and that each man's life is his own share of this common property. Either of these beliefs, if it does not destroy all sense of dependence upon God, leaves this sense imperfect and uncertain. Such a belief that man's life is his own is a falsity, and may well become a fatal falsity. It has a direct and a strong tendency to self-pride, self-trust, self-love; and these are the sources of all sin and all misery.

On the other hand, the belief of man's entire and

constant dependence upon God has, in all ages, been carried so far by some religious men, as to make man, in his best estate, only the passive instrument of God; and it prescribes to man, as his highest duty and his highest condition, to submit himself, without responsive action on his part, to the divine will and the divine action.

This too is a falsity; less mischievous, less dangerous than the other, but still a falsity. It is harmful in many ways. It prostrates the man before what is little else than a divine fatality. Its complete surrender of all selfhood paralyzes the will, and makes a healthy activity and a living interest in the duties of life impossible. Such may be its effect on those who succeed in this surrender of selfhood. To those who cannot succeed or who will not make the attempt, it suggests an excuse for the absence of all effort to grow better. It says to them: You are in the hands of God, let Him work his will upon you. Of yourself you are nothing. If it be His pleasure to make you good, He will do so, and you cannot resist Him. Nor can you if it be His pleasure to leave you the wicked man you are, or let you grow worse. If he who is good has nothing to do with his own goodness, surely he who is sinful has nothing to do with his own sinfulness. If God is all, and you are nothing, what have you to do but submit to your destiny?

Bring the two truths together, and we are safe. If we believe, and with every day strengthen our belief, that we are more than dependent upon God, that we live from Him at every moment of our lives, that our whole life and every thing which constitutes or belongs to life is given to us to be our own, and that freedom, as an element of that life, is given us to be our own, — freedom so to use the life thus given us that we may become His conscious, living, willing, and active instruments, or to turn away from Him and pervert and abuse our life, — then shall we be safe from the danger of supposing that we live from ourselves, and from the other danger of supposing our salvation from sin is a work wrought for us with no need of co-operation on our parts.

There are two ways in which a man may believe that his life is his own, which are exact opposites of each other. One is to believe that human life is self-originated and independent. The other is to believe that it is our own by the constant gift of God. The first is false, and the source of falsehood. The other is true, and a centre from which truth radiates. The apostolic command, "Work out your own salvation with fear and trembling, for it is God who worketh within you both to will and to do of his good pleasure," can be understood when the light of this truth is cast upon it. The word "for" in this passage might better be translated "because:" it indicates that what follows this word is the cause or reason for what precedes it. But is it not strange that we are required to do a work ourselves, *because* another does it all himself?

It is of the will of God and His good pleasure that all men should be saved. To that end He causes His own life to enter within us. It comes always and precisely so qualified and modified as to be in perfect adaptation to our state and our needs at that moment. Thereafter God ever worketh within us that His will and good pleasure may be carried into effect. Nothing that infinite love can prompt, or infinite wisdom devise, or infinite power do, to lead us to work out our salvation, is omitted; but this limitation remains for ever. We can work out our salvation only because He worketh within us to enable us to do this. And He can work within us only to enable us to work out our own salvation, — only to enable us to work with Him. He can do no more. All His providence, the constitution and ongoing of the universe, all the circumstances of our lives, all the influences which are caused or permitted to act upon us, the greatest and the least, all are governed by Him to this end. He can do no more. He cannot violate this law of human freedom, because this law is but the method by which infinite love and wisdom seek to impart to man the highest good and the greatest happiness. For this must ever be, the choice of good rather than evil, the love of goodness in itself, in and from our own actual and perfect freedom.

XIII. OUR LIFE OUR OWN AND YET GOD'S LIFE.

We cannot advance far in any just idea of God, without seeing that He must be alone and infinite. He alone can have life in Himself, and from Himself. All life but His must be derivative. We may perhaps suppose that man lives, or the world exists, because He commanded them to be; or, in other words, that He created them by His will out of nothing. This question has already been considered, and may be again; and the reasons more fully given for the statement that He does not create out of nothing, but from Himself. If He is infinite, there must be a sense in which He is All; for if there be any thing outside of Him and independent of Him, this something must be an addition to or more than infinitude. He creates from Himself; He creates by imparting Himself. He makes man to be a living man, by imparting to him life from His own Life. When this truth has full possession of the mind and is seen in all its clearness, then it will be seen and known that, at every instant of the life of every man, he owes the life of that instant, be it of the thought or of the will or of the body, to the life of God flowing into him at that instant.

This truth has no claim to novelty. When it is said, in the second chapter of Genesis, that "the Lord God formed man of the dust of the ground, and breathed into his nostrils the breath of life, and man became a living soul;" and when Paul says that "in Him we live and move and have our being," and the apostles in the Epistles use, as they do four times, the phrase that God is "all in all," it may seem to be implied that, in some way, we have our being because God imparts to us His being. But I would present this truth as a precise, exact, and absolute truth; for only when we hold it so, do all its momentous consequences follow. Then it would be impossible for us to live for a moment in the belief that we are independent of God. We should feel that He was our Father, as we could not otherwise. We should no longer believe that He gave us life by a solitary act at the beginning of our being; for we should know that He gave it to us then, and gives it to us ever since, and ever since as much as then. From such a conviction as this, not only the sense of dependence upon Him, but the constant thought of Him, and look to Him, and regard to His will, in all the hours of life, would necessarily derive new permanence and strength.

This is the good it would do. But can it do no harm? Yes: much harm, if this conviction stands alone in our minds. As has been already intimated, it might generate a feeling of fatality, a sense of constant subjection to overruling power. It would intensify our sense of dependence upon God to the suppression of all sense of dependence upon ourselves, or of responsibility for wrong-doing. It would give into the hands of God the whole work of our lives, leaving no part of it in our own hands. It would make of us the mere lifeless instruments, the mere channels of his will. Or it may exert a different influence, and give us over to enthusiastic self-conceit. We may think God has given us His own life, Himself; nor could we otherwise live. What more can we want, what more can He give? Are we not as God; are we not, each of us, a God to himself? From some feeling of this kind, not exactly defined in the consciousness, but very powerful, and never more powerful than now, a part of existing infidelity has arisen; and in its turn it asserts — often in different language — and confirms the fearful falsity that man is sufficient unto himself.

It has always been the great problem of religion to reconcile a perfect dependence upon God with the free will and free agency of man. They have not been reconciled. Religious men looked first and most at this dependence: anxious to fasten this belief in their minds, and conscious how it was dimmed and resisted by all worldly influences, they seemed to care for nothing else. They labored to impress upon their own minds, and upon their hearers, this truth in its fullest intensity. The Imitation of Christ, Law's Serious Call, and the Theologia Germanica — three most excellent and most useful books — carry this doctrine, so to call it, to the last extreme. The last-named expressly declares that the last good of human possibility is to be "the Lord's Hand;" and most earnest and pathetic eloquence is employed in the effort to make men seek to become mere unresisting instruments of God, through whom He can do his work with nothing of their work mingled with it.

This is an excess, but it is an excess of a good thing; and it is an excess which was justified and perhaps made necessary by the perfect antagonism between the truth and all that belongs to the merely natural man when that revolts against the spiritual. Hence the opposite extreme is far more common. The irreligious man considers religion as slavery.

He feels, and all these ascetic writings seem at least to justify his feeling, that religion claims not only his obedience, not merely the destruction of his freedom, but the suppression of himself. He will have none of it, that he may have full possession of all that belongs to his nature; and for that purpose he vindicates his freedom by denying his God. Generally, men have stood somewhere between these extremes, shifting their position from day to day, or from mood to mood; at one time inclining more to their sense of dependence upon God, and at another to their desire for freedom and their consciousness of it, always going to either by leaving the other, and never seeking to reconcile the two into unity.

And yet this can be done. They may be reconciled into a perfect unity. And this is done when we hold with equal clearness and certainty the two propositions that our life is God's life always and incessantly given to us, and that this life is so given that it becomes and is in us our own, — our own life, perfectly our own, and none the less always and constantly God's life in us. It is given to be perfectly our own, that our freedom in the use of it may be perfect; may be real, and not apparent only. It is still our Lord and Father who, in His infinite Love and Wisdom, so gives and so governs this inflowing life in every particular of every life, that all is done for man to induce him and to help him to use this freedom aright, but nothing to impair this freedom.

All our nature, all that belongs in any way to our natural faculties and affections, tells us that our life is our own. In this it does not tell a falsehood. But it tells us also that all these faculties and affections, with all the life they constitute, are self-originated; that they are ours by the inherent quality of our nature; that they are not from God, nor from any source other than ourselves; that they are not only in us, but from us. And this is not only a falsehood, but the worst of falsehoods.

Then the spiritual part of man comes in aid of the natural part of man. The higher accepts all of truth that the lower has to say. It accepts its assertion that our life is our own; but it rejects the falsehood that we live from ourselves. It recognizes the need that we should do our whole duty, earnestly and devotedly; and that we may so do our duty it knows that God has given us with our life a consciousness that our life is our own. But it also knows that this consciousness is not true nor trustworthy if it exists *alone* in the mind. It knows that from the beginning of human life there have been with all men intimations of the higher truth which consecrates this lower truth. And probably none have wholly escaped from the conviction that there is a higher power than we are, a Creator, a Father, who gave us life and governs our life. There may have been and there still are those in whom perfect and indurated worldliness appears to suppress every thought and every feeling which has not this world for its home; and all worldliness must do so in the degree in which it has an influence over our thoughts and affections. And there have been and are those who, in the pride of their self-ascribed reason and philosophy, have renounced all higher thought or feeling. But it may well be hoped that the light of the central truth of all religion has been sometimes able to penetrate even these dark and deep shadows.

It has been said that the natural faculties are given us to utilize this world, and the spiritual faculties are given us to prepare in this life for the unending life which follows this. But the spiritual faculties make this preparation, in part and a most important part, by using the natural faculties for this end.

XIV. WHAT IS THIS PREPARATION?

Again and again has it been said that this life is intended to be a preparation for another and an unending life. I would attempt to say what this preparation is, but to exhibit this in any fulness would require another volume. I can only give some indication of it in its most general form.

Heaven is God's Kingdom. There he reigns, and is known to reign. But how poor this language is, for it suggests only command and rule and sovereignty! They who are there recognize in its entireness the truth that their life is His life given to them to be their own. Their life and their happiness are measured by the completeness and the purity of their reception of His life and of His happiness. And His life is Love and Wisdom, in constant activity. They are able to receive more of this love in the measure in which they have put away from themselves self-love; for this is the one opposite to divine love, the one barrier to its reception. They are able to receive more of this wisdom in the measure in which they put away from themselves that belief in their self-sufficing independence of God, which is the centre of all falsehood. They see Him in all His goodness, in all His constant goodness to them and to all. They cannot but love Him supremely, and their love to Him gives birth to a love for others which casts off all selfishness as an abhorred thing. They see Him in all His providence; and they are wise, because the reception of His wisdom makes them able to understand His wisdom in its working, and to discern clearly what they must be and what they must do, that His will may be done, and His love carried into its fullest effect. They seek no more than this; but this they ever seek, and ever find in uses and activities as far beyond our imagination as they would be beyond our present powers. They look after their duty, He takes care of their happiness.

This is Heaven. And every hour of our life on earth offers us an opportunity to become more prepared to enter into this heaven. For, if we make a rightful use of the faculties given us for this purpose, in every hour we may resist and weaken some proclivity to sin or selfishness and error; we may strengthen our belief in God, and our trust in Him; and learn more and more to look upon every duty which He lays upon us, as His gift, given to us because every duty well done in obedience to Him, however humble it may be, bears us onward towards Him and His Heaven.

XV. HOW THE SPIRITUAL FACULTIES REGARD THE NATURAL FACULTIES.

The higher recognizes the lower, and all there is of truth or of good in the lower. And then it seeks to infuse into all this a new life, and set it on the way from earth to heaven. It bids the lower faculties and affections stand firmly on the earth, doing their whole work there; but with new activity and new success, because with a new heart and a new soul, with new motives, new ends, new strength, and new happiness. And it builds up a spiritual character by this activity and exercise and consequent growth and development of the spiritual faculties through the natural faculties.

In this we have an illustration of the relation between the higher and the lower, and of what the higher does with and for the lower. It is scrupulous to save all that is good and true below itself. It does not quench the smoking flax nor break the bruised reed. It does not say to the lower, Pass away into silence and inaction, and make room for me who am so much better than you. It ignores nothing that the lower does for the comfort of life or in art or philosophy or science. It accepts all this, and seeks for it, and earnestly promotes it.

For it knows that all the lower can do, anywhere and everywhere, is but a means towards an end which the higher cannot accomplish without the lower. In the far future it sees a boundless advance in the activity and fruitfulness of the lower faculties in all their vast diversity of action. And in this advance it sees new means and new methods for the accomplishment of its own work, which stands to the work of the lower faculties as the soul is to the body, as eternity to time, as heaven to earth. It gladly recognizes natural goodness, recognizes it *as* goodness, and rejoices in the belief that it is, in its own way and measure, a preparation for happiness hereafter. "In my father's house are many mansions."

There is a possible danger in dwelling too exclusively upon the belief that we live from God. It may tend to throw all our hope and all our duty upon Him, leaving nothing for us to do. But this danger is in these days very seldom incurred. It is far more necessary to consider the other danger, for that is actual and constant, — the danger that our sense of personality and freedom may exclude or overpower our belief in our dependence upon God. Safety from this peril is not to be found in weakening our sense of personality and freedom; for that cannot be too strong. Let this be doubted, let this sense grow obscure and uncertain, and in precisely the same measure all sense of duty and responsibility will become obscure and uncertain. And it ought to become so, because duty and responsibility rest most really and absolutely upon the *fact* that we possess our life, and all that constitutes our life, as our own, in every just sense of that phrase and under every aspect of the truth it expresses.

There is, perhaps inevitably, a tendency among thoughtful minds to consider the perfect sovereignty of God as implying some want of actual free agency with man. They know that freedom and duty must have an existence of some sort, but are disposed to regard these as so far inconsistent with divine omnipotence and omniscience, that our consciousness of them seems to be not so much the true consciousness of a positive fact, as a permitted illusion, — permitted to us and good for us, that by it we may be induced to bring our conduct into conformity with divine law. But this consciousness is not a permitted illusion. It is a consciousness of the fact upon which rest heaven and earth. Our life is our own. It is not *from* ourselves: it is all, always, and instantly, *from* God and given to us; but it is given us to be our own, and to constitute our selfhood. This giving of life from God to become most actually our own is the foundation upon which rests all the work of God, in the creation, preservation, and government of the universe, — natural and spiritual, — and of all things therein.

Geology and all the sciences which tell us of the vast periods consumed in the growth and development of the material universe; history and ethnology, and whatever speaks to us of the growth and development of the human race through the immeasurable past, — all help us to see the infinite patience of the Lord. To see it; but do they help us to understand it, to know why the Omnipotent accomplishes only very slowly, with long delay, and with a long series of short steps of progress, results which it would seem that an almighty fiat might have brought into being at once? No. To understand this at all, we must cast upon it the light of the great truths, that the end of creation is a constantly growing heaven of the human race, and that men can be prepared for heaven only in their freedom, only by their voluntary co-operation with their Father's working. The day may come when this truth, taken in connection with the correspondence of all things of the material worlds with all things of the spirit, may solve the mystery of those unnumbered ages during which earth, this and other earths, were and are prepared to be the dwelling-places of human beings.

Even now we may catch a glimpse of the truth which explains the slow and gradual advance of man from the rudest beginnings to the present hour, and his extreme imperfection at this hour, and his utter inferiority, as yet, to any true and high standard of excellence. For we may see that all the advances of mankind, all the steps of their progress, all the accumulated good of this progress, constitute a work in which man does his part, in the freedom which God gives him that he may use the constantly inflowing life of God, in doing this work as of himself.

If we can see this truth distinctly, and accept it unreservedly, we shall see that, were not man's life his own, he could not love God in perfect freedom and by his own choice. We shall see that this gift of life to be our own is that which lifts man above the brutes, and makes man to be the summit and crown of creation, and builds him into God's own image and likeness, and makes it possible for him to have that pure and perfect love of God, which gives him the capacity of receiving that highest happiness of which created beings can be capable. And if we admit that it must be the desire of infinite love to create beings to whom it may be possible for Him to give this happiness, then we shall see that God could not but give life to man to be his own; that he cannot withhold or qualify this gift, or disregard it in the slightest degree in any part of his government of man. And man has the power of continuing or becoming bad, because, if this power were taken from him, with it he would lose the power of becoming good.

Then we shall be no longer troubled to account for the existence of evil. We shall not be able, and we shall make no effort to fathom the mystery of the infinite; but so much as this we know. Man must be free; not in name only or in appearance only, but most actually and most absolutely; and it is just so that he is free, because life, and freedom as an element of life, are given to him to be his own. That is not freedom which is limited or controlled in its exercise or in the direction it would pursue. Therefore, freedom to do good implies of necessity freedom to do evil; or, in other words, if evil could not be done, good *must* be done, and what must be done cannot be done in freedom. And because this freedom is most real it may be abused; and it is abused. Hence moral evil. And if the externals and surroundings of man are made for him they must be in adaptation to him, and must be what he makes it necessary that they should be that this adaptation may be preserved. And thus moral evil calls physical evil into being.

XVI. OF THE PROVIDENCE OF GOD.

And what is there in this sense of absolute and constant freedom to weaken our sense of dependence upon God? Nothing, providing we hold as firmly to the other truth, that every thing whatever of our life comes from God, and is continually His gift; always and at every instant the gift of infinite love and infinite wisdom, always so measured and so modified as our highest interests require.

It may be true that one may believe that his life is incessantly given him of God, and is given him to be his own; and nevertheless so far regard himself as independent of God, as to be incapable of humility and exposed to the fearful danger of self-sufficiency. This may happen thus. If one believed God gave to him at birth life from Himself, to be for ever his own, it is easy to see how he might come to feel that this life, whatever its origin, when once his own, made him as independent and self-sufficing as if his life were self-originating. And even if he came nearer to the truth, and believed that this life was not given him at once and for eternity, but always and at every moment, he might still feel that when thus given, if given to be his own, it made him equally independent and self-sufficing as if given at the beginning. It is, then, to guard against this error and this danger, that a third truth comes in. The first two give the facts, that human life is divine life given to man, and that it is so given to him that it becomes in him his own life. The third gives the reason for this. It is that man may in the exercise of his freedom become prepared for happiness. But, that this end may be reached, it is not enough that divine life be given to man to be his own. But it is constantly so given and constantly so modified in its influence, and all external things are so arranged, that all that is possible may be done for man, to lead him, guide him, and induce him to exercise his freedom in such wise as to approach the end for which this freedom is given.

Divine life is, in itself, perfect love and perfect wisdom. And they are with us always. They do not leave the man to whom they have given and in whom they preserve existence. From the first moment of that existence they are doing, and never cease to do all that Omnipotence can do, to help him to happiness. Whatever is, is under divine government, the greatest and the smallest alike; for all are so perfectly. Nothing can happen under this government but for an *end:* this end, as it is that which infinite love seeks, must be the highest happiness of the creatures it has caused to be. Therefore we may know that every event, whether it be of history on the largest scale, or of the moments in the life of the individual, must happen for his highest good; and this means his eternal good. This highest and eternal good requires that his freedom be always given, always respected. Whatever happens must happen for his highest good, and because it is suited to produce that good. But it can produce that good, or any good, only so far as he is willing that it should.

Here we find the limit, and the only limit, to what God can do for man. It is the limit imposed by His love and wisdom upon His omnipotence. Any thing and every thing is always done to help man to *choose* right and not wrong, in his own freedom. More God cannot do, because He cannot cease to be perfect love and perfect wisdom.

But is this freedom never to be taken away when the hour of happiness comes: is it not to be controlled for the sake of that happiness, and to prevent its loss? The answer is, Never. And if the question then is asked, What security can there be in the happiness of heaven, what certainty that it will not fall before the same dangers which imperil and impair our happiness on earth? the answer to this question can have no meaning, unless it gives a glimpse, if even that be possible, of the very essence of heavenly happiness.

We live here to prepare for heaven. This preparation is made only in and by and through our freedom. And this preparation is effectual when such a character is built up within us, that, while our freedom will always be perfect, we shall always exercise our freedom in choosing good. Not because we *must*, but because we *will*. And this is heavenly happiness. Even there may be alternations of state; for freedom is there, or it would not be heaven. Even there selfishness may grow too strong, and threaten to cast its chill and darkness on the love of God and of the neighbor. Even there it may be necessary that we be reminded that two ways are open before us, and only one of them is our Father's way. But they who are there are there because they are so reborn into newness of heart and life, and their character is governed and determined by such principles and motives, that it is only necessary for them to see these two ways, to make it certain that they choose the better.

This, then, may be the conception of the highest man,—of him who has grown into the full stature of a man; that is, of an angel. His consciousness and certainty that his life is in very fact and truth his own are strong, but far stronger is his gratitude towards the infinite love which made it possible for him to resist and overcome his proclivities to sin and selfishness and degradation. And at the very foundation of every thought and affection lies the certainty that each and every one of them, and all that he has and all that he is, flows to him constantly from this constant love. He is there the instrument of God, the *hand* of God, but yet a person and himself. He is the living instrument, who, in the freedom and the strength given to him to be his own, accepts and uses all his means and capacities for doing his Father's work. We can neither imagine his grateful joy when he remembers that he is so constituted, so led, and so supported, to the very end that he may be capable of receiving this happiness from his Father's Love, nor his shuddering horror at the thought of being deserted by his Father and left to himself.

XVII. REVELATION.

Revelation has already been defined, or rather described, as Truth taught by the Lord directly to mankind. What is meant by this word "directly"? We know what sensation is, and we know, though not so precisely, what the mind can acquire by thought about sensations. The acquisitions thus made may be very large and diversified; perhaps, within their own scope, unlimited in extent and variety. But there is a limit which they cannot pass; for they can never go beyond things appertaining to this material universe, and to the life of man while he is in the material body. For thoughts and ideas which transcend this limit, knowledge must come to us by other means than by sensation. All thoughts come to us from God, because all thought and all life come from Him. But all of those which I call natural thoughts and ideas come to us by the instrumentality of the material universe, and of those faculties which are adapted to make the utmost possible use of this universe. The knowledge which we have of spiritual things can not come to us by this instrumentality. It comes directly. It comes by Revelation.

There are many, perhaps indefinitely various ways in which it may be given to man. If we remember that Revelation is a giving by the Infinite of knowledge concerning infinite things to finite beings, — and that nothing can be given that cannot be received, — and that ages and races differ indefinitely in their capacity of reception, we may be sure of two things: one, that Revelation must be multiform; the other, that it must be progressive. It must be multiform, because it must be adapted in form and manner to the varied receptivity of ages and races. It must be progressive, because that truth is always given, and only that which can be usefully given. And the purpose and the effect upon human character of the spiritual truth given by Revelation must always be to enlarge human receptivity for spiritual truth.

Between sensation as the foundation of all action of the natural faculties and Revelation as the foundation of all action of the spiritual faculties, there are many points of analogy.

Sensation teaches at first simple things, easily received; and, as the senses grow in power by exercise, the instruction they give grows in extent and in definiteness. But always what it teaches must be received rationally, and can only yield its best fruits when the faculties fitted to deal with it do their work wisely, and advance, step by step, in the conclusions they draw from what the senses present to them.

So Revelation, which began we believe when man began, has been received in a vast variety of ways and of forms, but has been on the whole advancing and progressive. So, too, it has always been given to the spiritual faculties; for, if man had not possessed these faculties, Revelation could neither have been given to him nor received by him. Moreover, it became what these faculties by their wise or their unwise action upon revelation made of it. And when these faculties became perverted, their comprehension of the truths of revelation became perverted, and their presentation of them distorted. For the spiritual faculties, when perverted, are just as capable of spiritual error and falsehood as they are of seeing and stating spiritual truth, when they are not perverted.

XVIII. THE SUCCESSION OF REVELATIONS.

Diversified as have been the forms and methods as well as the measures of revelation of spiritual truth, they have been alike in one respect. They have always been given through some use of the natural faculties. They have always been given for the spiritual faculties to lay hold of and profit by. They have always been used, and they have always been abused. So far as they have been used, they have developed and invigorated the spiritual faculties of man, and raised them into the possibility of apprehending higher and further revelation; and then this has been given. So far as they have been abused, their truths have become perverted into falsities; and they have been corrupted until they became the apparent causes of, or incitements to, or excuses for, every kind and every degree of wrong-doing.

Over all this Divine Providence has watched; always respecting the spiritual freedom of man's ownhood, always preserving it by equilibrating the spiritual influences which come to him, and always doing for man all that the maintenance of this freedom permitted. So far as the gifts of revelation have been perverted and abused, the evil results of this abuse have been mitigated and counteracted, until they passed away to return no more. And, so far as the gifts of revelation have been rightly used, the good results of this use have remained as abiding means of improving human character. Hence there has been an advance — by waves which sometimes seemed to retreat, but on the whole did advance — in the condition of mankind; a gradual growth in the capacity of receiving revelation and making a good use of it; and with this an advance, in successive steps, of the revelations which give men spiritual truth.

Then, as the human race advanced under the influence of revelation, and as the truths revealed were higher, their perversions even into entire falsities grew less dreadful; and the wrong-doing perpetrated under the name of religion, although still the worst of wrong-doing, was mitigated in its character. Many ages have passed away since any religion could be so falsified as to persuade its followers that they pleased their Gods when they cast their children into a huge brazen figure, flaming with heat. And many ages have passed since any religion could impel its votaries to the public practice of unutterable lust and cruelty. Those things have passed away, not to return.

Not to return in those forms. But a perversion of Christian doctrine and motive led to the infernal and indescribable cruelties practised upon the Waldenses, and in the Netherlands, and by the Inquisition generally, and not by that alone. And the "faith alone" of Calvin and Luther hastened into premature ripeness, and bore its fruits, and by its fruits showed what it was, in the abominations of antinomianism in Munster. But all that too has passed away, not to return.

From the beginning of human existence there have been revelations. For man has always been immortal, has always lived on earth to prepare here for a happy immortality, and could never make this preparation but by the use of spiritual faculties given to him for this end; and these faculties could have no foundation to rest upon, no beginning to proceed from, and no activity whatever, except by making use of what was given to them by revelation. For the natural faculties alone could not offer to them even a suggestion, or impart to them the first movement of spiritual life.

These revelations have been exceedingly diverse in way and manner, and form and measure; for in many ways and by various means have truths concerning God and the spiritual world come down to man from God through the heavens. And this descent of truth has always been a revelation. There are four of these of which I would speak more particularly.

In ages far before historic times, there was a church founded among men. They who were its members differed, I believe, exceedingly from ourselves in constitution and in character. Revelations were made to them in accordance with their peculiar constitution and character. What these were, I can describe only by saying something of the correspondence between the world of spirit and the world of matter.

XIX. CORRESPONDENCE.

God creates from Himself, by effluence from Himself. What He thus creates is nearer to Him, or farther from Him. What He thus creates nearest to Himself (we may call it, if it will help our understanding of the case, the sphere of being), this nearest sphere is the instrument by which He thus creates the next; and so on successively until the last and lowest, earth, this and other earths, are so created. Between each sphere thus created, and that above it by which as an instrument it is created, there is perfect correspondence. Each therefore corresponds to all, and all to their Creator.

A volume would not suffice to exhibit either the laws or the facts of this correspondence in detail. It may be impossible for me to say any thing intelligible about it, in a page or two. But I must make the attempt. This correspondence has been the foundation of all human speech; for in all languages words have expressed things above the senses only by rising into that meaning from their original lower sensuous meaning. It has given to poetry all the force and beauty it has ever had; and by means of poetry, and because recognized when so used as a fanciful and pleasing unreality, even by this use of this correspondence thoughts have been suggested and motives prompted by poetry, which otherwise the world might not have known. But the time has come for the recognition of it as a universal fact, and the study of it as a science; and as a science which must lie at the foundation of all philosophical religion. I would therefore, if I could, offer to my readers a sketch, which must be brief and scanty and utterly imperfect, of some of the elementary principles and conclusions of this science.

Whatever belongs to man, and is not of his body, may be referred either to what he loves, desires, or feels, or else to what he thinks, intends, and believes; in other words, all things of the world within man refer either to the Will, or else to the Understanding. So all things of the world without man, which by correspondence represent the things within man, refer either to the things of the will or to those of the understanding.

The infinite love of God flows into the human will, and there constitutes whatever of affection the man has. This love flows down below the world of spirit into the world of matter. The primal and most general form which it takes there is Heat Already science is rapidly advancing towards the conclusion that all the forces of nature are but forms and modifications of one force, which, in nature, may best be called Heat. It will reach this conclusion, but will not stop there; for natural science will be led by spiritual science to know that Heat is but the form which Divine Love puts on, when it comes down into nature and operates there. And there are indefinitely diversified forms and effects of heat in the world without man, and indefinitely diversified forms and effects of love in the world within man; and these correspond, all to all, each to each, with precise and scientific exactness.

What is said of the relation between Love and Heat may be applied, with little change, to the relation between Wisdom and Light. This word wisdom I must take to express every thing which, proceeding from its infinite source in the Divine Being, fills and animates every thing of the understanding, or of the intellectual side of his nature, in man. And whatever light is or does in external nature, that this influent wisdom does in the realms of thought; and these two correspond, all to all, and each to each, with precise and scientific exactness.

XX. THE TEST OF CORRESPONDENCE.

These details might be carried indefinitely farther; but what has been said will suffice, at least for a sample. It will suggest this view to some readers. There is a sort of resemblance or analogy between heat and affection, and between light and truth, which common language continually expresses. Swedenborg, not content with this in its common-sense form and measure, lays hold of it as a foundation on which his strong scientific tendency leads him, and his vivid imagination enables him, to build up a sort of system which he calls a science. But passing beyond the bounds of a moderate, common-sense use of an obvious analogy, he plunges into folly and mysticism. So some readers will think; so perhaps they cannot but think. But to others it may occur as worth inquiry, whether it is not possible that these simple, common-sense notions are but the dim recognitions of a greater truth, and are themselves true only because they are the most obvious applications of what is in fact a universal law. And that Swedenborg, the eminent mathematician and scientist which all men called him once, has succeeded in learning and in teaching the principles of a true science.

It ought to be plain, but I fear it is not, that the science itself offers the only test of its own truth. If, when patiently and honestly studied, it offers no fruits to the student, there is for him the end of it. But if, when carefully examined in its principles and in the application of them to the facts of mind and nature, it is seen to solve many difficult questions, and to throw a strong light into dark corners of both worlds, and reveal unsuspected treasures in them, surely it may be deemed a science, and one of much value. It will indeed cover the whole of both worlds, and reconcile them. It will make the outer world the mirror, the expression of that which is within. And it will teach us to find words and figures to express truths of the inner world, which would otherwise be unexpressed and inexpressible.

XXI. THE ANCIENT CHURCHES.

The most ancient church above spoken of learned spiritual truth through these correspondences. It was common among them to have their spiritual senses open, and they lived in a communication with the spiritual world which was afterwards lost. They heard persons speak, and saw persons and things, who were only in that world. They saw the correspondence of natural things with spiritual things, and in natural things saw the spiritual things to which they corresponded and which they therefore represented.

This was the infancy not of the human race, but of churches among men. When this state passed away, the knowledge of correspondence remained, not as matter of perception, but by memory and tradition; and as the ages passed this became obscure and perverted; but it was never wholly lost among men, and has been the permanent foundation of spiritual beliefs which have always prevailed among the nations, and have had an immense diversity of aspect and of character.

In a subsequent age, distant from the beginning, but so distant from our own times that history has but dim intimations of it, these traditions of correspondences were gathered into form. It may have become later a written Word, but at first we have no reason to believe that writing was known. It was then delivered over to memory and systematic tradition. Modern inquiries into the remains of antiquity bear interesting testimony to the systematic, faithful, and successful efforts to preserve the accuracy of tradition, by means of a body of men thoroughly trained in the art of accurately remembering, preserving, and teaching truths held sacred.

XXII. THE BIBLE.

We have grouped all these modes of revelation together, regarding them as substantially one. And now I come to the second. It was that which gave us what we call the Old Testament.

The Wisdom of God took possession, so to speak, of the writers of the books which compose that Book, took possession of their pens, and of their hands which held those pens, and of their minds and memories, and of the thoughts they had and the words they had. Through these it made a Revelation of itself, as perfect as the instruments it must employ permitted. And it is divine throughout. In its letter it comes down to men in all states and conditions to do for them the good for which all revelation is given, — to lift them above a merely natural condition, and into one which may be receptive of spiritual and abiding happiness.

To natural men, and to men in the lowest degradation of sin and darkness, it comes on its errand of mercy; and it says to them all that can be said to such as they are, in the literal sense. But it comes also to men in higher, and in all higher, even the highest states, on the same errand. It teaches them what they are in their own selfhood, marred and degraded as that is, and the absolute antagonism between this and good and happiness and God, and by what means they may escape from sin and suffering, and depart from it farther and farther for ever and for ever. This it does in much of the literal sense, and throughout in the spiritual sense which pervades the whole Word.

It was for this purpose, — that the literal sense might be by correspondence an adequate expression of the spiritual sense, — that the Wisdom of God took possession of the writers, and suspended for a time their personality and freedom. A part of the earlier Word above mentioned, which was composed in distant ages wholly by correspondence, and was without literal truth, was placed as an introduction to this Word, and extends into the eleventh chapter of Genesis. Under the form of the creation of the external universe is told the story of man's internal creation, of the birth and growth of his spiritual nature, or of his regeneration.

The remainder of that Word consists of religious songs, of prophecies, and of the history of the Jewish nation. That this history might, in its letter, be an adequate expression of spiritual truth, the Jews were led through their strange and eventful history, in such wise that the mere narration of those events suffices for the most part as the form of spiritual truth. Not always, however: in some instances — taken altogether, only a very few — this was impossible, and statements were made not historically accurate, because it was necessary so to give spiritual truth under natural forms. The impossibilities lately so much commented upon by Colenso and others are instances of this.

The Jews were a chosen people, — chosen not for their goodness, but for their peculiar character. They were chosen as the most merely natural people then or ever existing. Hence in the literal sense there is little reference to another life, and they could learn little of another life. Their God was presented to them as one of terror, of vindictive wrath; and pestilence, famine, and war were the instruments he loved to employ. Such at least is the aspect of the presentation of God to them in a large part of the literal sense of the Old Testament.

And that is the aspect of the Word of God however uttered, or of Divine Truth, to all in whom the Jewish nature predominates; and this means to all merely natural men who are down at the bottom of the scale, and to this nature wherever it exists, and therefore to all men until they rise high enough to see this Truth under another aspect. It is to them who are in this low condition, commandment, coercive law: "Thou shalt not." It puts on this aspect that it may reach and help them; for "The fear of the Lord is the beginning of Wisdom." But obedience to the law gradually improves the character, and opens the mind to see it under a different aspect. The commandments become laws of love, and these laws become promises of a state in which law no longer needs to be coercive. Obedience will still be rendered, no longer by fear, but by gratitude for the goodness which gives these laws as guides to happiness.

XXIII. THE FIRST CHRISTIAN REVELATION

The Bible is the expression of Infinite Wisdom: in its letter it is the lowest expression. But as character advances, as with regeneration a new spirit is born within us, we become capable of seeing higher senses to this Word; and above the highest angels who rejoice in the brightest sunshine of heaven, there still rises infinite wisdom, the Word that in the beginning was with God and was God.

And then, in the fulness of time, this Infinite and Divine Word came down to earth in a new way. Assuming from Mary a human nature, He stood upon earth, a man among men.

I am sure that I had better not attempt to give, in the brief space I could here devote to it, any statement of what I should call the Doctrine of the Lord: not because it would be feeble and imperfect; for that any statement I could make of this transcendent doctrine, any explanation I could offer of this transcendent Fact, would be, but because it would be so imperfect as to be unintelligible, unless I gave to it an altogether disproportioned space of this little essay. Let me, then, go at once to the character of this, the first Christian revelation. It is made by the Gospels and by the book which bears the name of Revelation. They, like the books of the Old Testament, were written by Inspiration, and in accordance with the correspondence between spiritual things and natural things. In their literal sense the Divine Wisdom is in its fulness and its power; and this sense will remain for ever, the lowest steps of the ladder which resting upon earth rises to heaven, and through heaven to its source.

Among the words recorded of our Lord is the promise that He will come again "in the clouds of heaven." He, the Infinite, who came to earth in the human nature which he assumed, will come again to earth, and this time He will come "in the clouds of heaven." What does this mean, and what can be this coming? The science of correspondence may give us the answer to this question. If it cannot, nothing else can.

XXIV. THE SECOND CHRISTIAN REVELATION.

All things of the external world correspond either to what is of the will or what is of the understanding, because these two constitute the world within. This has been said already. I add now that water corresponds to what I may call truth, having no word of wider meaning, but intending to include all things of the understanding which belong to actual life, as water does to the earth. Clouds are water lifted up from the earth by the sun, to the intent that it may fall down again to fertilize the fields and support life. By inspiration, infinite wisdom lifted up from the earth thoughts, knowledges, words, and facts which belonged there, and used them as its instrument or means, by which this truth might descend upon the fields of spiritual life, and fertilize them and support spiritual life. Hence clouds correspond to and represent the letter, or literal sense, of this Word; and they bear this meaning throughout the Scriptures. When our Lord declared that His second coming should be in the clouds of heaven, he declared that it should be in the literal sense of His Word. He has so come. He has come in and by a revelation of that spiritual sense of Scripture which lies within the literal sense.

This revelation He made through his servant, his agent for this purpose, Emanuel Swedenborg. The general facts of this man's life are now well known. I need refer to them but very briefly. He was the son of a Swedish Bishop of high repute, and was born in 1688. Educated as well as was then possible, possessing remarkable intellectual power, with an earnest thirst for knowledge and habits of devoted industry, he acquired in early life a high position among scientific men, and produced successively works of science and philosophy which have always been acknowledged as having great merit. About the year 1745, when fifty-seven years old, his spiritual senses were opened, and, abandoning natural science, he devoted himself for the remainder of his life, twenty-seven years, to the acquirement and promulgation of spiritual science, — of the science of religion.

His spiritual senses were opened. The material body is formed of material substance. It has already been said, that there is but one substance, and that is Divine. Flowing forth from the Infinite, and reaching its lowest place, there, through its adaptation to the mind and the senses, and through the agency of the mind and the senses, it becomes matter, and of this the body and the bodily organs of sense are formed. But within the material body is the soul; and this too is an organism, living in and acting through a spiritual body, which, with its sensories, is formed of spiritual substance; and this is the same substance from the Divine, in a higher form. These two forms or modes of substance correspond together. Therefore it is that the spiritual body fills and animates the material body, and lives in it and acts through it; for whenever we see, or hear, or feel, it is our spiritual body which sees and hears and feels through the material body; which loses all life and sense when the spiritual body leaves it at death.

This material body has two functions. By one, it is the instrument by which the senses of the spiritual body may have, through the material senses, cognizance of the things of this world. By the other, while it clothes the spiritual body with the material body, it is a barrier between the senses of the spiritual body and the things of the spiritual world. That is to say, while we live in a material body we are cognizant of the material world. But because the material body is in the way, we are not then cognizant of the spiritual world. The material body loses both of these functions at death, because it then ceases to clothe the spiritual body. Thereafter the spiritual body through its sensories has cognizance of the things of the spiritual world, and has no longer any cognizance of the things of the material world. Either of these functions of matter may be suspended during life in this world. The material body by sleep, or disease, may cease to be an instrument through which the spiritual senses take cognizance of the things of the material world. And also it may happen, that the material body ceases to be a barrier, preventing the spiritual senses from taking cognizance of things of the spiritual world. Either of these functions may be suspended partially or entirely. The cause which prevents the material body from being an adequate instrument of the spiritual body may operate upon some only of the senses or of the limbs, and its operation may be complete or imperfect. Just so, if the barrier be removed, this removal may be partial or entire, it may be complete or imperfect. Only one sense, or only two, of the spiritual body may be enabled to take cognizance of the things of the spiritual world, as sight or hearing or touch only; and they may have this cognizance dimly and imperfectly, or clearly and perfectly. In Swedenborg's case the barrier was entirely removed; almost as perfectly as it is removed by death. He was in the spiritual world, not always or continuously during these twenty-seven years, but nearly so; a spirit among spirits, almost as completely as when he finally left the material body.

If the main purpose of this revelation was that our Lord might come "in the clouds of heaven," or that he might come and disclose that infinite wisdom which lies within the letter of His Word, the question may be asked why, for this purpose, the spiritual senses of Swedenborg were opened. The answer to this question must come from the science of Correspondences.

XXV. A DIFFERENCE BETWEEN THIS MATERIAL WORLD AND THE SPIRITUAL WORLD.

This correspondence is, in the most general form, a correspondence between the spiritual and the natural. Hence this lower outer world corresponds to the world within man. But this correspondence is general only, and not specific; it is a correspondence of the external universe to man's internal, but not an exact correspondence of the external world surrounding each individual to his internal. The external world is, so to speak, the common resultant of all the forces which form the whole world within men, and they give to the world without him its form and aspect and correspondence. To this extent, therefore, it is a mirror and exposition of the internal world; and this it is in all its forms and all their changes, in all its laws and their effects.

Far more than this is true of the spiritual external world; or of the world of which every one has cognizance after death, and they who still live here have cognizance in the degree in which their spiritual senses are open. In that world this correspondence is not general only, but most special and specific. There, every thing immediately around a man is in exact correspondence with that which is in his affections or his thoughts.

The reason of this difference is this. We live in this world to change our characters, — to change them for the better, radically; and we can do this only by effort and struggle. Hence we need and have an external world, which being vested, so to speak, or ultimated in a stubborn and indurated matter, resists us. We wish it something else than it is, for it does not respond to our feelings or desires. But it does not yield to us. It reacts against us. Sometimes we cannot even hope to make it responsive to our desires, and we make no effort. Sometimes we may hope to do this, and we strive to do it, and our efforts are more or less successful. This man, for example, finds his external circumstances painfully opposed to his wishes. He wishes to be elsewhere than where he is, or to be richer, or more healthy; and, if he thinks success possible, he labors to produce the change. But whether his labors effect this or not depends upon whether it is best for him that they should. Seldom are they perfectly successful; or, if they are, disappointment and unsatisfied desire are sure to come in some other directions. All this is because these external circumstances which oppose his desires so much are just the instruments by which Infinite Love gives him the lessons he needs, and subjects him to the discipline he requires, in order that he may, by a radical change in his desires, — that is, in his character, — become capable of true and permanent happiness.

It is not so in the other world. We live here, in time, to determine our characters, to determine them for eternity. We die when they are so determined. Whether we do or do not profit by the lessons or the discipline of love, depends upon ourselves. But when the perfect wisdom which is one with that perfect love sees that we have reached the point where it is better that this conflict shall cease, it does cease: the man dies; that is, he leaves this material world and goes into the spiritual world; that is, again, he loses all sensuous perception or cognizance of the things of this material world, and comes into a perfect sensuous perception and cognizance of the things of the spiritual world.

At once the scene is changed, and the law is changed which determines the relation of the external world to the internal. The character is determined. The man no longer needs a world outside of him which shall react forcibly against his inner man, and he has no such world. The external world now acts only for and with the internal world, and does not react forcibly against it. It is needed only for the manifestation, the exercise and development of character, and not for its radical change; because the time when that was possible has passed by; or, rather, because that could take place only *in time.* Therefore it is perfectly responsive to the inner world; not only in general and as a whole, but specifically and particularly. And therefore, each and every man finds in all about him, which is *not* him, a ready instrument for, and an exact representation of all that *is* in him.

This is true in its whole extent and without qualification, only in heaven; that is, only with those in whom character is determined to goodness. Few, however, go thither from this world at once, and without needing some discipline to liberate their goodness from its enemies within them. They need a discipline suited to this purpose. And those who do not use aright the means of salvation given them go not up to heaven. To spirits in the early stages after death, and to the spiritually dead, the external spiritual world still responds to their internal specifically and precisely; but now to their states and their needs, and not always to their wishes.

XXVI. SWEDENBORG.

Now let it be remembered that, by the laws of correspondence, whatever appears outwardly and objectively represents and mirrors something within the soul; and then that the Scriptures were written in exact conformity with the laws of correspondence; and it may be seen why it was necessary that he whose mission it was to learn the laws of this new science and apply them to the letter of Scripture, and teach men how to do so, found in this opening of his spiritual senses a most potent instrument for the instruction which he needed in the nature and working of correspondence. Nor is this all. Thus intromitted into the spiritual world while still living in the natural world, and able to go there and hear and see and learn, and then to return hither and put into an abiding form the information he thus acquired, he was also able to present this information to men on earth in intelligible forms, by the help of the illustration afforded by these correspondences. So he could make us see and feel what goodness was and what it led to, and what evil was and what it led to, as we could not otherwise.

This man brought to the service assigned him his vigorous powers, highly cultivated, exercised, and disciplined; and he received instruction and information by means never employed as to any other man, in any thing approaching the same degree. But he was not inspired. Between the Word of God and his writings there is the infinite distance which separates that Word from all human work. As well as he could, and we may believe as well as any man could, he understood what he was thus taught, and gives to us. And we may well believe, too, that, strong as he was, cultivated and prepared and taught as he was, and employed as he was, he was guarded against important error. But between all this and the inspiration and authority which belong to the Word, there is an infinite difference and an infinite distance.

It may be asked whether Swedenborg has given us a full, explicit, and precise spiritual sense of the Holy Scriptures throughout the Word. And, if not, how can what he gives us be called a Revelation of this spiritual sense of the Scriptures?

He has made no such Revelation. But he has declared the existence of the science of correspondence. He has taught the principles of this science and the laws of being on which it rests. He has disclosed the spiritual sense of a part of the Scriptures.

If this be all, it may be called a most imperfect Revelation. It certainly is so in one sense; and in the same sense every Revelation of the Infinite to the finite must be so. But as a Revelation it is complete and sufficient for its purpose. Like every other that has been made, or ever will be made, it is delivered to man, to be made use of by him. He gives us the laws and principles of this science, and applications of them which show that there is such a science, and how its laws and principles may be applied.

It remains for us and for all future generations to do our and their part. We may accept the Revelation or reject it. We may study these laws and principles, and endeavor to comprehend them clearly, and make farther and wider applications of them, or we may refuse to do this. We may undertake this work, and do much or little of it, and we may do that well or ill.

All this is given up to our freedom, our choice, our action. But if we look into this Revelation until we discern evidence of its truth, and then endeavor to see these truths yet more clearly, and make them our own by honest and earnest study of them, not only for knowledge, but for life, — for all who do this the Word of God will become the Sun of our souls, pouring its heat and its light into our winter and our midnight.

If such a man as Swedenborg tells us such new and important truth, why is it not more rapidly and more widely received? My answer is, it cannot, and never can be, received by any but those who are favorably disposed towards it.

I know perfectly well how readers who are not of the New Church, if I have any such, would understand this statement, and what would be their answer. They would say to me, Of course it is just as you say. This good and very able man did perceive and has expressed some important truths in religious philosophy; and some of these — all which could stand the test of rational investigation — have been widely received, and do now exert much influence on the best religious thought. But he was an imaginative man, devoted to his abstruse studies, leading a secluded life, and not protected from the vagaries of enthusiasm and the delusions of imagination, either by domestic ties and cares, or by the work and responsibility of professional or official life. This last he held until he threw it aside at the age of fifty-nine, and with it he threw aside that which had been the barrier between him and mere illusion. Then in his seclusion, concentrating all his powers in one direction and indulging all the tendencies of his character, his *subjective* notions became to him at last as *objective* things. His profound conviction and his vivid imagination enabled him to impart to his account of these things much fascination for persons who are like him, imaginative and enthusiastic. His own convictions he gives to them. They believe in him. And only such persons can believe in him.

Why would this answer be made to me? Is it because what he has said of super-natural things, or what he has related from his spiritual experience, is in itself irrational? Not in the least. Certainly I have never known or heard of any one who had investigated these things sufficiently to form a definite and well grounded opinion of them, and who formed that opinion. It is because saying any thing whatever of super-natural facts, or speaking of spiritual experience, seems in itself irrational, contrary to reason. And it is entirely contrary to merely natural reason,—so contrary, that, when this reason is listened to as a competent judge of these questions, it cannot but be deemed irrational. And we live in days when the spiritual faculties have little power or authority, while the natural faculties have vast power and almost complete authority. The explanation of Swedenborg's state and condition, which I have supposed to be made by those who do not receive his narrations of spiritual facts and events as true, is exceedingly plausible. It is indeed entirely satisfactory to those whose habits of thought and faith make it impossible for them to believe any definite statements concerning spiritual existences, although they are able to receive some, perhaps much, of the new truth taught by him. It is well for them that so plausible an explanation can be made; for otherwise they would be compelled to reject him altogether, and thus lose what they now gain and may profit by.

But they who believe these statements of Swedenborg do not regard them as isolated and independent facts. They constitute a material part of a system of truth which is most comprehensive and reaches to the inmost depths of thought. They are in harmony with the principles of that system; and the general facts he states might be inferred, without his statement, from the principles and laws of that system. And while these principles and laws explain those facts, the facts in their turn illustrate those laws and principles, and make plain to us their meaning, working, and results.

In the light of that system, we see that there is another world; that there must be a world of spirit and of eternity, to account for this world of matter and of time. And we see, too, that the world of spirit must be such a world as that which he describes. Because his system is a whole, and between all its parts there is harmony and congruity, all support all; they illustrate and prove each other. We believe in the laws of being which he teaches, all the more because in his narratives of spiritual existence we can see and understand how these laws operate and what they effect. We believe his narratives, the more because they so perfectly conform to the laws of being which he states. Insanity is not so coherent, and mere imagination not so rational. In a word these narratives must be true if those laws exist and operate, and those laws must exist and operate if these narratives are true. Nor is this reasoning in a vicious circle. How many of the most important conclusions of natural science have been reached and are now held, because the proof of the facts on which they rest is strengthened by their harmony with the laws which explain them, and which in their turn they illustrate.

All this is nothing to habitual naturalistic belief, which is the prevailing belief of the age. When I say this, I may be referred to

XXVII. SPIRITISM.

It may be said that the rapid growth and present prevalence of Spiritism is a proof that I am mistaken as to the character and tendency of the age, because it proves that the spiritual faculties are not only awake and active, but eager to seize, with indiscriminating hunger, upon whatever is offered them. To this I reply that Spiritism (I think this is a more appropriate word than Spiritualism), so far as I have been able, not without some effort, to understand it, is the most purely natural belief that has ever been held among men, and that its quick and wide reception is a cogent proof of the present feebleness and inaction of the spiritual faculties.

I have used both the words "external" and "natural" in preceding pages, but the last much more frequently. They are, however, when used in this connection, synonymous and interchangeable. The peculiarity, I might almost say the definition, of the natural faculties, is, that they are employed exclusively about the external world. There is in the other life an external world just as much as there is in this world. It is like this, but with a difference. It is perfectly like this in this respect, — that there are faculties perfectly adapted to it. They are the same which I have called natural faculties, and have said to be adapted to this external world. They belong to the body and to external life here; and because man has a body and external life there, to which these external faculties are suited and are necessary, he has these faculties there. They are there, when subordinated to and wisely used by the spiritual faculties, instruments of immeasurable value. They are there, when they suppress the spiritual faculties or rule over them, just as disastrous in their influence and effect as they are here. And those who go from this world with a purely natural character retain it there, and find an external world adapted to it, and continue to be for ever just as natural as they were in this world.

Most true it is that, but for the spiritual faculties

which man possesses, he could not have the first thought of life after death. But it is equally true that the natural faculties may seize hold of this truth, make it their own, use it or abuse it as their own. The spiritual sensuous faculties necessarily awake at death. If only this were needed for man's instruction, if a knowledge that there was another world, and that man had a body and sensories which would make it a home for him, were all that was required to make man wise, all men would be made wise by death. But death, of itself, or an awakening after death, does nothing to make men wiser or sillier, better or worse. They rise in that world just what they were in this world. They may reject every supersensual truth, and every religious truth, and exclude every religious affection, just as effectually and thoroughly there as here.

The external or sensuous spiritual faculties may be called spiritual in so far as they belong to the spirit. But when they are not elevated by and governed by the higher spiritual faculties, the man himself is just as much an external man, or a merely natural man, as he was before death substituted the sensuous spiritual faculties for the sensuous natural faculties which were open here.

Such persons are very numerous in the other world· and they are drawn by the attraction of affinity to such persons here. And, when both parties have a certain impressibility of the spiritual sensories, those who live here become mediums; and those who live there talk through them; and, so far as I have been able to learn, nearly all they say is pure naturalism. They may personate — for what I know, they may be weak enough to believe that they are — departed friends of those who seek them, or the great and good men whose names they take, and whose authority they try to give to words which such men could not utter unless their intellects were wofully clouded.

It is among the peculiarities of the other world which distinguish it from this, that there thoughts and feelings are known; that is, as a general thing, they are perceptible and are seen to be what they are. The good do not desire it to be otherwise: the bad cannot prevent it. The spirits who come to the mediums, and address themselves to those who seek them, and may be supposed to be for the most part like them, find little difficulty in knowing what is passing in the visitors' minds, and in making use of it to make their communications such as will give them possession and control of their visitors. But good spirits would shrink from this with aversion and dread. They know too well the value of freedom to wish to impair it in others. These evil spirits would, if they could, possess the very bodies of their suppliants, and they do take possession of their minds. They may, if they see that it would help them, simulate enough of religion to answer their purpose. But, for the most part, — and I do not know enough of their writings to justify me in speaking more strongly, — they teach only the falsities of naturalism, and inspire only corresponding affections.

Each of the words "natural" and "spiritual" is used in two senses. By natural we may mean the faculties or qualities which belong to man by birth and are adapted to earthly life, and are also adapted to be instruments of higher faculties in the preparation for a higher life, and in the enjoyment and use of the external world in another life. And when we speak of natural faculties we may mean this. Or we may mean these faculties or qualities when they are confined to external life, and, renouncing all subordination to the higher faculties, either suppress or pervert them; and then and thus they fill the heart and life with worldliness and selfishness.

So we may mean by spiritual only what belongs to man as a spirit, whether it be good or bad. Or we may mean that condition which results from the rightful employment of the higher faculties.

So far as the naturalism of these days is merely the negation of any thing of life after death, spiritism opposes it, and finds no difficulty in the relations of Swedenborg. But, so far as naturalism clings to mere earthliness of character, — or, to say the same thing in other words, so far as it renounces God, and with God all true religion, admitting none but what is consistent with self-worship, — it is utterly opposed to the doctrines taught by Swedenborg. His doctrines "give to God the glory" of all the goodness which man has from Him; for them our Lord and Saviour is our God; to them the Scriptures are the Word of God; they command self-denial, and purity of life, affection, and thought; and they give to the marriage relation sanctity, purity, and permanence. To all this naturalism is antagonistic, and the naturalism of spiritism almost entirely so. I remember some spiritist writings — among them a book by an English lady, entitled "From Matter to Spirit" — which were certainly religious. But I have met with very few of this kind.

The mere knowledge that there is another life is not in itself of the slightest value. Indeed it may do harm by taking off the fetters of fear from those who would only abuse their recovered freedom. Hence Divine Providence has permitted the grave to cast such deep shadows over the whole earth. Nor have former revelations made the certainty of a future life — by giving the ground and manner of its being as well as the fact of its being — such that most men were able to go farther than cling to a hope. Spiritism is no revelation; and, if it does more in reference to the spiritual world, there is too much of it which, if it comes thence and is not mere illusion, comes up from the depths of self-love and self-worship, and drags its votaries thither.

XXVIII. WHO RECEIVE THE LATEST REVELATION.

Let me return to my remark that none can receive the doctrines of Swedenborg who are not favorably disposed towards them. There is not, there never was, and never will be, any religious truth ever given to man, which was not and will not be so given that, while he who loves it may be convinced of its truth on rational grounds, they who have no love for it may reject it on grounds which seem to them equally rational.

This is a most important law, or fact; and the reason for it is not far to seek. We need but look to the love of God. We need but remember that Infinite Love must desire to do for man the best that can be done, to give to man the best that can be received. And surely it must not be difficult to see that the best happiness that man can have must consist in his choice of all goodness in his own perfect freedom; and therefore this freedom is an essential element in the life given to man to be his own, and cannot be wholly lost out of that life without his ceasing to be human. Infinite wisdom is in God one with infinite love; and in their perfect union they make Him to be one God. Wisdom in action becomes order; and perfect wisdom becomes perfect order, divine order: and it is this and this only which imposes limits to Omnipotence; but these limits are never passed. Hence, while all is done that Omnipotence can do to lead men to choose goodness instead of evil, truth instead of falsity, nothing more is done by Omnipotence. Nothing more can be done by Divine Omnipotence, for the very reason that it is the omnipotence of perfect love and perfect wisdom.

While in this truth we have the key to the whole providence of God, in nothing is its explanation clearer or more necessary than in the questions which spring from the form and character of all revelations. Never was there one, and never can be, which is not either received and rightly used, or else rejected or perverted and falsified, just as he to whom it is given CHOOSES.

To speak only of the Old Testament, — its obscurity, and utter failure to compel a reception of truths which they who choose to receive find to be inexpressible blessings, are thus accounted for. And so of the Gospels. Their seeming contradictions, indefinite statements, and mere hints and suggestions of profound truths, have, on the one hand, led to much doubt and denial, and, on the other, to all manner of perversions of religious doctrines. But they have never hindered those who came to them as to wells of living waters, to be cleansed and healed, and to learn the way of life. Our Lord spoke in parables, and "without a parable spake he not unto them." He himself gives the reason for this. And, when this reason is understood, it is seen that it was this infinite mercy which clothed the divine truth he uttered in the clouds of parable; and that from the same mercy He has now come in these clouds. But still they are clouds. And still clouds and darkness must be about his throne.

This law will help us to understand many of the peculiarities in the writings of Swedenborg, which — while wholly uninspired, and forming no part of and no substitute for the Word of God — are the instruments of the latest revelation. They make him a hard author for many to read. They do not present doctrine in a clear and analytic form, beginning from elementary principles, advancing along a plain and easy way, and leading the mind, as by a pleasant journey, to results which cannot but be reached.

It is the infinite desire of God to enlarge man's freedom; to make it pure, perfect, constant. And His successive revelations advance in this direction. That given to the Jews was sustained by tremendous instruments, — by plague, famine, war, and captivity. But these were so adjusted, that even they could not compel a constant and universal obedience, although they did promote and cause much of that external obedience which alone was and is possible to the Jewish character, in the Jews of old, or in us.

Then came the first Christian revelation. This was supported by miracles. But they were no longer miracles of seeming wrath. They were miracles of love; not more so in fact than the sanctions of the Jewish law, but that the love from which they flowed was manifest in them. They were miracles not of coercion, but of conviction. And who needs to be reminded how little power they have exerted when they were opposed by the inertia of indifference, or, more actively, by the hostility of sin and self-love?

Then comes the second, the consummating, Christian revelation. And now not only are all miracles withheld, but the means and method of this revelation are such, that, while it gives itself like sunlight to the soul whose windows are opened, it is wholly unable to penetrate the barriers interposed by any want of inclination for the good to which this truth leads. And this is what I mean, when I say that the truths of this revelation can be received only by those who are favorably disposed towards them. Never let it be forgotten that truth alone is only like the light of the coldest winter day, that is utterly incapable of awakening into life the dead earth which it floods with brightness. Well is it for them who seek truth for something else than the good of life to which it leads, that their eyes are holden.

XXIX. THE WORD OF GOD CANNOT PASS AWAY.

The word of God, and whatsoever belongs to it, is infinite and eternal. Plague, famine, war, and captivity are still the sanctions of divine truth. But it is now the pestilence which works within the soul, and smites that, if truth be rejected, with leprosy and palsy and spiritual death. It is "a famine not of bread nor a thirst for water, but for hearing the words of God." It is war, our own war with the enemies of our souls; and it is captivity when they have made us captive.

When our Lord said, "Believe me for the very works' sake," His words expressed, as all the words of Him who spake not as man speaks ever expressed, an infinite and eternal truth. In these days, also, we must believe for the very works' sake,—yes, for the sake of the same miracles; nor can we believe, nor do we ever believe wholly and heartily in Him by the power of whose Word they are wrought, on any other ground.

But now we believe for the sake of these miracles when their vestments of earth are cast off, and they are transfigured into miracles of the spirit,—miracles worked in our spirits by the Spirit of God. Then we do indeed believe in Him; for we are sure that He has fed our famine with the bread of life, and slaked our thirst with the water of life. He has lifted us from the bed whereon the fever of selfishness had laid us prostrate, and made us whole that we might minister to Him; He has healed our sicknesses, He has strengthened our weakness, He has opened our eyes and our ears, and we no longer walk lamely along the paths of life: we are raised from spiritual death to spiritual life. I say we are raised, that we do all this, that we believe thus: I mean only that this possibility is placed before us. And, while we know that we have as yet only touched the hem of His garment, we may perhaps have reason to believe that some power has come forth from Him. And, if it has, it fills us with unclouded faith, a hope to which there is no limit, and a humility which casts Self prostrate at His feet.

XXX. FUTURE REVELATIONS.

I have called this a consummating revelation. It may be that many successive revelations may be needed, and therefore given, to explain and complete this. But there would seem to be a valid reason for believing that no one will ever stand in the same relation to this, which this holds to all by which it has been preceded.

Infinite wisdom, and this is absolute and perfect wisdom, expresses itself in the Holy Scriptures. They are the Word of God. This infinite wisdom is hidden under the letter. But it is all within the letter. It is there not conventionally and arbitrarily, but by the same law which governs the creation of the universe and all the acts of infinite power. For this law is the law of correspondence. By this law the wisdom of God expresses itself in the works of God. They are its image and its continent, and they are its exponent to all who can discern their meaning. This is true still more emphatically of His Word. As a man's thoughts dwell in his words and are clothed by them, and as we understand his thoughts in the degree in which his words are their adequate expression and we understand them, precisely so, but in an infinite degree, the whole wisdom of God dwells in His Word and is clothed by it. He has made the letter the adequate expression of that wisdom. The only thing that remains is that we should understand His word.

By this latest revelation He has given to men the means of doing this. As yet we can do it only in a very external way, and most imperfectly. But He gave to his servant Swedenborg instruction which enabled him to lay the foundations of the science of religion and religious philosophy, by giving to us their most important principles. These principles are now within reach of the spiritual faculties of man; and his natural faculties, if the lower be subordinated to the higher, will all cooperate in the comprehension, application, and use of these principles.

This progress must be gradual and slow, but it will be eternal. The science of religion, like every true science, must begin with its elements, and advance with successive steps. This advance will never end. As men grow wiser in their understanding of the Word, it will be to them what it is now to men who have become angels. It will become to men the Word of God as that exists in the heavens. But there two are all degrees of this wisdom, and in each degree a constant advance. This advance, as it is constant, so it will be eternal; and yet it will never ascend to its source, for that is the inmost of the mind of God. And this must ever be infinitely above the highest elevation to which created and finite intellects can ascend.

XXXI. HE COMES WITH POWER.

When our Lord said that He would at some future time be seen to come in the clouds of heaven, He also said that He would come "with power."

When has He not come with power? The Creator of all things, infinite, and therefore All in All, All in all things always and everywhere, in what grain of sand, in what blossom that opens under his sunshine and his rain, in what insect that draws the breath of life and of enjoyment, does He not come with power? But He said, "'They shall *see* the Son of man coming in the clouds of heaven with great power." This is the prophecy that is now fulfilled. Men may now *see* this power which is put forth in all that is, but has hitherto been hidden under a veil that was always thick and but imperfectly transparent, and now is lifted.

There is still another and a higher meaning to this prophecy, and this too has been fulfilled.

The Word of God is His instrument. "All things were made by Him." But most especially did this Word, this infinite wisdom, take on the form of human words, that as the written Word it might come to men, to cleanse and purify and elevate them, and lead them to happiness and to Himself. And how has this power of the Divine Word been now increased? It has become more than it was, in the measure in which spirit is more than matter, eternity more than time. It is as if the arm of the Lord had been made bare. Still, however, as in the past, so always the putting forth of this power must be governed by that wisdom which is one with the love that fills it. Still it must respect man's freedom; that freedom which it seeks to free from all obstruction, to enlarge and to establish. Still, men who crucify Him by denial, indifference, or disobedience, will continue to rend his outer garment — the literal sense of His Word — into pieces, each holding that fragment which his own depraved character can most easily reduce to become its instrument; and still will the Lord withhold his inner garment, one and without seam, from them who would rend that also, or profane or pervert it, or cast it idly away, as men would cast away a pearl when they had no knowledge of its value and no eyes for its beauty.

If this latest revelation be what it declares itself to be, it must be true that some reject it because it is above their capacity or their desire for goodness. But it is not true, it is very far indeed from true, that none reject it and none are ignorant of it, but those who are thus low in character. A very different reason may operate with the majority of those who know nothing of this revelation, or who, knowing something, reject what they know. This reason is simply that it does not suit them. The blind man, opening his eyes for the first time on a noon-day sun, would not be more dazzled by the intolerable light, than we should be were our eyes opened to the wonderful adaptation to our needs of the means provided for our salvation; and the special and precise adaptation of these means to the special and precise needs and possibilities of every individual. What each one requires, and can make use of to help him on his upward path, must depend greatly on his inherited character, and on the circumstances about him, and on all the habits of his education, thought, and life. These must determine what is best for him, both as to the spiritual truth presented to him, and as to the manner of its presentation.

Hence we may explain the marvellous failure of the vast efforts of Christian missionaries among the heathen. Ages ago Catholic priests went among the same peoples, and "converted" them by thousands. But the conversion was almost wholly external, extending but little beyond name and ritual, and reaching to forms only and not to faith. And such a conversion might do some good, and could not do much harm. But now the effort is to make the heathen discard old doctrines and accept new ones. And it is nothing less than wonderful that efforts which are so earnest and so persistent should be so ineffectual. There may be other reasons for this, but there is reason enough in the simple fact that the religion of the missionaries *does not suit their hearers*, is not adapted to perform for them the only use which religion can perform, — that of lifting their thoughts and affections to our Father in heaven, and to the truths and laws He has given His children as the rules of life.

Very various are these systems of religious truth; for they need to be so, that they may be adapted to the various states of those to whom they are given. No Christian man can doubt that Christianity is, in itself, better than heathenism, but it is better for some and not for all; and there is not and never was a heathenism which, with all its follies and falsities, had not in itself the means of salvation; and it seems only a reasonable inference from all we see and learn, that this day as many persons find and use these means as there are those who find them in Christianity, and that they use them as effectually.

We who have faith in this latest revelation must of course believe that it is in advance of all that have come before it. But we do not think that we, personally, are in advance of all that are outside our boundaries; and God forbid that we should be so blind as not to see in some of those who know nothing of our doctrines, — or, knowing them, cannot see their truth, — purity, charity, living faith, and excellence of motive and of conduct, before which we bow with reverence, and in which we would find examples and incentives.

XXXII. AND WITH GREAT GLORY.

Yes, with all glory; for the glory of the whole earth, of the whole universe, of matter and of spirit, is now His. He "has taken to Himself his great power, and reigns." He may now be seen to be the king of all things, and, more, the life of all things. His reign is a reign of love, not merely because it is animated and governed by love, but because it is love, His love, which creates all things for the reception of itself, and flowing into them becomes their life.

All valid and enduring advance in human character is an advance in the knowledge of God. So, too, all advance in successive revelations lies in their enlargement of the means of knowing Him. To the Jews it was given to know that He was One and sovereign. They were to be the custodians and publishers of this transcendent truth. But for the hardness of their hearts they were permitted to believe that He was especially their God, and they especially His people, and the objects of his favor. When our Lord walked on earth, a man with men, he revealed, both by His words and His acts, far more of the true nature and actual working of the Father within Him, that, seeing Him, they might know the Father. And this latest revelation explains His words and acts, and gives those central truths concerning Him from which all light may flow forth, and those organizing truths which will some day build up in heavenly forms the structures of human belief, concerning Him who may now be known as our God, our Creator, our Redeemer, our Lord and our Saviour.

Already the disciple that Jesus loved had said that "God is Love." Beautiful the light, enduring the strength, unspeakable the consolation, these words have given where light and strength and consolation were most needed, although none have understood these words as meaning more than that

God was perfectly loving. But we know now what is the essence of love: it is the desire to give of its own to others, to give of itself, to give itself. This desire is infinite in God. It leads him to create those who may receive his gifts; to create them with capacities of reception of every degree, and then to fill those capacities with Himself.

He gives to all men, always, all the truth and all the good affections, yea, all the life and being, that they have. His is a divine giving, and therefore a most actual and perfect giving. It is so complete and perfect, that man can own nothing so perfectly as he owns himself; and he owns most that which most essentially constitutes himself. How feeble is his hold upon natural possessions! They are his but for a moment, and even then but imperfectly his. But his spiritual possessions, the thoughts of his understanding and the affections of his will and his very life, are perfectly his own. And yet all the while they flow into him from the Father of his being, by a constant flow and as a constant gift. His life is thus his own, constituting him himself, because the infinite desire of God to give Himself to man leads Him to make man such that he possesses the power of receiving the divine life, and holding it as his own. Because this human life is thus, most really and permanently, his own, he is not a lesser

God, he is not a part of God: he is a man, he is himself; and he never can be lost in God, never can be merged in the Divine Being, never can lose his own identity. The doctrine of final absorption into God prevailed in some of the old religions after they had become perverted, and lingers in them still as they linger upon earth. And on examination we should see evidence of its greater prevalence among Christians in past and present ages than would at first be supposed. But it is totally false. Man never can be *united* (made one) with God: but he can be *conjoined* with Him, this conjunction being of two, each of whom has his own separate identity.

Thus the infinite All-Father has a boundless universe of creatures to whom He may give Himself, and thus satisfy the infinite desire of His infinite love. And then this love labors infinitely to make this reception of Himself larger, more entire and perfect. All below man can have but a reception of the divine life limited in each, at once and from the beginning without change, by the limitations of his nature. Not so with man. His reception may grow for ever, more perfect, more pure, more unperverted; because there is given to him, in the life given him to be his own, the capacity of gradually putting away from him all that resists the reception of divine life, or that causes the perversion of it when received.

And now we may see what is the constant end of Divine Providence in all its working. It is to make the conjunction of man with Himself more perfect; and that it may be more perfect, to help man to put away from himself all that opposes or impairs this conjunction. And because this can be done by man only as the self-love of his inherited nature is resisted and put away by the exertion of the strength given him to be his own, therefore is it the perpetual effort of Divine Providence to strengthen man's Ownhood of his life, and to vivify that Ownhood, that human life, with good from Himself, that through it man may become more and more perfectly conjoined with God.

No one can know better than I do how poor and dim a presentation of a great truth my words must give. But I write them in the hope that they may suggest to some minds what may expand in their minds into a truth, and, germinating there, grow and scatter seed-truth widely abroad. I am sure only of this: The latest revelation offers truths and principles which promise to give to man a knowledge of the laws of his being and of his relation to God, — of the relation of the Infinite to the Finite. It gives new motives for seeking, as well as new means for finding truth, when that is sought to make us better; a new guidance in the darkest and most difficult paths of life, new comfort in its desolations, new strength in our weakness. It breaks the seals of the Book, written within and without and sealed with seven seals, which no man has hitherto been able to open. And therefore I believe that it will gradually, — it may be very slowly, so utterly does it oppose man's unregenerate nature, — but it will surely advance in its power and in its influence, until, in its own time, it becomes what the sun is in unclouded noon. And the sign and the effect of its establishment in the hearts of men will be, that the whole earth will be filled with the glory of the Lord.

Cambridge: Press of John Wilson and Son.

PINK AND WHITE TYRANNY.

BY MRS. HARRIET BEECHER STOWE,

Author of "Uncle Tom's Cabin," "The Minister's Wooing," &c.

With Illustrations. 1 vol. 16mo. Price $1.50.

This vivid and pointed novel of modern society presents to us in a story which will not be called exaggerated, some of those phases of life around us which are none the less dangerous because they are called contemptible. The extravagance of the newly rich, who have never learned the use of money, and the failure of the substitutes by which people who live by sensation try to supply the place of honor and religion, have never been portrayed more precisely. At the same time Mrs. Stowe does justice to that sex which is not enough remembered in the discussion of the wrongs of Woman. For she describes, as no one else can describe, the tyranny under which a loyal and chivalrous gentleman suffers most terribly. The pen, which more than any other quickened the public heart till the black slavery of centuries was broken, will render a service not less considerable, if it so wake the conscience of men and women that pink and white tyranny of women over men shall be impossible.

ROBERTS BROTHERS, *Publishers,*
BOSTON

THE GREAT RELIGIOUS BOOKS OF THE DAY.

ECCE HOMO.
ECCE DEUS.

Although it is now two years since the publication of "Ecce Homo," and one year since "Ecce Deus" appeared, the sale of these extraordinary and remarkable books continues quite as large as ever. Some of the ablest and most cultivated minds in the world have been devoted to a critical analysis of them.

The foremost man in England, the Right Honorable W. E. Gladstone, has just published a book devoted entirely to a review of "Ecce Homo," in which he uses the following language: —

"To me it appears that each page of the book breathes out, as it proceeds, what we may call an air, which grows musical by degrees, and which, becoming more distinct even as it swells, takes form, as in due time we find, in the articulate conclusion, 'Surely, this is the Son of God; surely, this is the King of Heaven.'"

Of "Ecce Deus," which may be considered the complement of "Ecce Homo," there are almost as many admirers, the sale of both books being nearly alike.

Both volumes bound uniformly Sold separately. Price of each, $1.50.

Prof. Ingraham's Works.

THE PRINCE OF THE HOUSE OF DAVID; or, Three Years in the Holy City.

THE PILLAR OF FIRE; or, Israel in Bondage.

THE THRONE OF DAVID; from the Consecration of the Shepherd of Bethlehem to the Rebellion of Prince Absalom.

The extraordinary interest evinced in these books, from the date of their publication to the present time, has in no wise abated. The demand for them is still as large as ever.

In three volumes, 12mo, cloth, gilt, with illustrations. Sold separately. Price of each, $2.00.

The Heaven Series.

HEAVEN OUR HOME. We have no Saviour but Jesus, and no Home but Heaven.

MEET FOR HEAVEN. A State of Grace upon Earth the only Preparation for a State of Glory in Heaven.

LIFE IN HEAVEN. There Faith is changed into Sight, and Hope is passed into Blissful Fruition.

From Rev. Samuel L. Tuttle, Assistant Secretary of the American Bible Society

"I wish that every Christian person could have the perusal of these writings. I can never be sufficiently thankful to him who wrote them for the service that he has rendered to me and all others. They have given *form and substance to every thing revealed in the Scriptures respecting our heavenly home of love*, and they have done not a little to invest it with the most powerful attractions to my heart. Since I have enjoyed the privilege of following the thought of their author, I have felt that there was *a reality* in all these things which I have never felt before; and I find myself often thanking God for putting it into the heart of a poor worm of the dust to spread such glorious representations before our race, all of whom stand in need of such a rest."

In three volumes, 16mo. Sold separately. Price of each, $1.25.

Mailed, post-paid, to any address, on receipt of the price by the publishers

www.ingramcontent.com/pod-product-compliance
Lightning Source LLC
LaVergne TN
LVHW011224110826
845150LV00006B/1539

* 9 7 8 1 4 2 5 5 1 6 1 2 3 *